统计与大数据"十三五"规划教材立项项目
数据科学与统计系列规划教材

人工智能专业英语

An English Course of Artificial Intelligence

张强华 司爱侠 ◎ 编著

人民邮电出版社
北京

图书在版编目（CIP）数据

人工智能专业英语：附全套音频 / 张强华，司爱侠编著. — 北京：人民邮电出版社，2021.7（2023.8重印）
数据科学与统计系列规划教材
ISBN 978-7-115-56089-6

Ⅰ．①人… Ⅱ．①张… ②司… Ⅲ．①人工智能－英语－教材 Ⅳ．①TP18

中国版本图书馆CIP数据核字(2021)第039334号

内 容 提 要

本书选材广泛，主要包括人工智能概述、现实世界中人工智能的应用、人工智能的类型、人工智能的利与弊、人工智能中的知识表示、人工智能中的推理、人工智能中的搜索算法、机器学习算法、专家系统、人工智能中的模糊逻辑及其应用、有监督机器学习与无监督机器学习、现实世界中机器学习的应用、人工神经网络、深度学习、模式识别、面部识别、人工智能软件开发工具、用于人工智能的编程语言、人工智能即服务、人工智能在云计算中的作用等内容。

本书内容与课堂教学的各个环节紧密切合，对备课、教学、复习及考试等教学环节有一定的指导作用。

本书既可作为高等院校本科、高职高专人工智能相关课程的教材，也可供从业人员自学使用。

◆ 编　　著　张强华　司爱侠
　　责任编辑　孙燕燕
　　责任印制　李　东　胡　南

◆ 人民邮电出版社出版发行　北京市丰台区成寿寺路11号
　　邮编 100164　电子邮件 315@ptpress.com.cn
　　网址 https://www.ptpress.com.cn
　　固安县铭成印刷有限公司印刷

◆ 开本：800×1000　1/16
　　印张：14　　　　　　　　　2021年7月第1版
　　字数：294千字　　　　　　2023年8月河北第2次印刷

定价：49.80元

读者服务热线：(010)81055256　印装质量热线：(010)81055316
反盗版热线：(010)81055315
广告经营许可证：京东市监广登字 20170147 号

前　言

人工智能正在引领这个时代的技术发展，为许多行业赋能。近年来，我国许多高校纷纷开设了人工智能专业，培养市场急需的专业人才。由于人工智能行业发展迅速，从业人员必须掌握许多新技术、新方法，对其专业英语的要求也比较高。因此，具备相关职业技能并精通专业外语的人员往往能够赢得竞争优势，成为职场中不可或缺的核心人才。

为了贯彻落实二十大精神，编者编写了本书。本书的特点如下。

（1）选材全面，内容严谨。本书对人工智能概述、现实世界中人工智能的应用、人工智能的类型、人工智能的利与弊、人工智能中的知识表示、人工智能中的推理、人工智能中的搜索算法、机器学习算法、专家系统、人工智能中的模糊逻辑及其应用、有监督机器学习与无监督机器学习等内容进行了介绍，并对课文素材进行了细致推敲与加工，使其更好地满足教学需求。

（2）体例新颖，适合教学。本书各单元包含以下部分：课文——包括选材广泛、风格多样、切合实际的两篇专业文章；单词——给出课文中出现的新词，读者由此可以积累人工智能专业的基本词汇；词组——给出课文中的常用词组；缩略语——给出课文中出现的、业内人士必须掌握的缩略语；习题——包括针对课文内容的练习、词汇练习、翻译练习和填空练习；参考译文——便于读者对照理解课文，提高翻译能力。

（3）教学资源丰富，教学支持完善。本书配有PPT、参考答案、音频文件和教学大纲等资源。另外，书中的习题数量适中，题型丰富，难易搭配，便于教师组织教学。

在使用本书的过程中，有任何问题，读者都可以通过电子邮件与我们交流，我们一定会及时答复。邮件标题请注明姓名及"索取人民邮电出版社人工智能英语参考资料"字样。我们的E-mail地址为zqh3882355@sina.com和zqh3882355@163.com。选用本书的教师也可到人邮教育社区（www.ryjiaoyu.com）免费下载相关教辅资源或到人邮数据科学与统计教学QQ交流群咨询：1056931673（仅限教师身份）。

编　者

目 录

Unit 1　Artificial Intelligence ········· 1

Text A　Overview of Artificial Intelligence ···· 1
　New Words ············· 5
　Phrases ·············· 8
　Abbreviations ············ 9
Text A　参考译文　人工智能概述 ······ 9
Text B　Artificial Intelligence Applications in the Real World ········· 13
　New Words ············· 16
　Phrases ·············· 18
　Abbreviations ············ 19
Text B　参考译文　现实世界中人工智能的应用 ············· 19
　Exercises ·············· 22

Unit 2　Types, Pros and Cons of Artificial Intelligence ········ 27

Text A　Types of Artificial Intelligence ···· 27
　New Words ············· 29
　Phrases ·············· 31
　Abbreviations ············ 31
Text A　参考译文　人工智能的类型 ······ 32
Text B　Pros and Cons of Artificial Intelligence ············· 34
　New Words ············· 37
　Phrases ·············· 38
Text B　参考译文　人工智能的利与弊 ··· 38
　Exercises ·············· 41

Unit 3　Knowledge Representation and Reasoning ········· 45

Text A　Knowledge Representation in Artificial Intelligence ········ 45
　New Words ············· 49
　Phrases ·············· 51
　Abbreviations ············ 51
Text A　参考译文　人工智能中的知识表示 ············· 51
Text B　Reasoning in Artificial Intelligence ············· 54
　New Words ············· 58
　Phrases ·············· 59

Text B 参考译文 人工智能中的推理 … 59

Exercises … 62

Unit 4 Algorithms in Artificial Intelligence … 66

Text A Search Algorithms in Artificial Intelligence … 66

New Words … 69

Phrases … 70

Abbreviations … 71

Text A 参考译文 人工智能中的搜索算法 … 72

Text B Machine Learning Algorithms … 74

New Words … 78

Phrases … 80

Text B 参考译文 机器学习算法 … 80

Exercises … 83

Unit 5 Expert System and Fuzzy Logic … 88

Text A Expert System … 88

New Words … 92

Phrases … 93

Abbreviations … 94

Text A 参考译文 专家系统 … 94

Text B Fuzzy Logic in Artificial Intelligence and Its Applications … 98

New Words … 101

Phrases … 102

Abbreviations … 102

Text B 参考译文 人工智能中的模糊逻辑及其应用 … 103

Exercises … 105

Unit 6 Machine Learning … 109

Text A Supervised vs. Unsupervised Machine Learning … 109

New Words … 112

Phrases … 114

Text A 参考译文 有监督机器学习与无监督机器学习 … 114

Text B Machine Learning Applications in the Real World … 117

New Words … 121

Phrases … 122

Abbreviations … 123

Text B 参考译文 现实世界中机器学习的应用 … 123

Exercises … 126

Unit 7 Artificial Neural Network ········ 130

Text A Artificial Neural Network ········ 130
New Words ········ 135
Phrases ········ 137
Text A 参考译文 人工神经网络 ········ 138
Text B Deep Learning ········ 141
New Words ········ 144
Phrases ········ 145
Text B 参考译文 深度学习 ········ 145
Exercises ········ 148

Unit 8 Pattern Recognition ········ 152

Text A Pattern Recognition ········ 152
New Words ········ 156
Phrases ········ 158
Abbreviations ········ 159
Text A 参考译文 模式识别 ········ 159
Text B Facial Recognition ········ 162
New Words ········ 166
Phrases ········ 166
Abbreviations ········ 167
Text B 参考译文 面部识别 ········ 167
Exercises ········ 170

Unit 9 Artificial Intelligence Software Development ········ 174

Text A Artificial Intelligence Software Development Tools ········ 174
New Words ········ 178
Phrases ········ 180
Abbreviations ········ 180
Text A 参考译文 人工智能软件开发工具 ········ 181
Text B Programming Languages for Artificial Intelligence ········ 184
New Words ········ 188
Phrases ········ 189
Abbreviations ········ 189
Text B 参考译文 用于人工智能的编程语言 ········ 189
Exercises ········ 192

Unit 10 New Technology of Artificial Intelligence ········ 197

Text A Artificial Intelligence as a Service ········ 197
New Words ········ 200
Phrases ········ 201

Abbreviations ……………………… 201
Text A 参考译文　人工智能即服务 …… 202
Text B　The Role of Artificial Intelligence in Cloud Computing ……………… 204
New Words ………………………… 207

Phrases ……………………………… 208
Abbreviations ……………………… 209
Text B 参考译文　人工智能在云计算中的作用 ………………………… 209
Exercises …………………………… 211

Unit 1
Artificial Intelligence

Text A
Overview of Artificial Intelligence

扫码听课文

1. Intelligence

1.1 What Is Intelligence

Intelligence is the ability of a system to calculate, reason, perceive relationships and analogies, learn from experience, store and retrieve information from memory, solve problems, comprehend complex ideas, classify, generalize, and adapt new situations.

1.2 What Is Intelligence Composed of

1.2.1 Reasoning

Reasoning is the set of processes that enables us to provide basis for judgement, making decisions, and prediction. There are broadly two types of reasoning (see Table 1-1):

Table 1-1　　　　　　　　　　　Types of Reasoning

Inductive Reasoning	Deductive Reasoning
It conducts specific observations to make broad general statements	It starts with a general statement and examines the possibilities to reach a specific, logical conclusion
Even if all of the premises are true in a statement, inductive reasoning allows for the conclusion to be false	If something is true of a class of things in general, it is also true for all members of that class

1.2.2 Learning

Learning is the activity of gaining knowledge or skill by studying, practising, being taught, or experiencing something. Learning enhances the awareness of the subjects of the study.

The ability of learning is possessed by humans, some animals, and AI-enabled systems. Learning is categorized as:

(1) Auditory learning: It is learning by listening. For example, students listening to recorded audio lectures.

(2) Episodic learning: To learn by remembering sequences of events that one has witnessed or experienced. This is linear and orderly.

(3) Motor learning: It is learning by precise movement of muscles. For example, picking objects, writing, etc.

(4) Observational learning: To learn by watching and imitating others. For example, child tries to learn by mimicking her parents.

(5) Perceptual learning: It is learning to recognize stimuli that one has seen before. For example, identifying and classifying objects and situations.

(6) Relational learning: It involves learning to differentiate among various stimuli on the basis of relational properties, rather than absolute properties.

(7) Spatial learning: It is learning through visual stimuli such as images, colors, maps, etc. For example, a person can create a roadmap in mind before actually following the road.

(8) Stimulus-response learning: It is learning to perform a particular behavior when a certain stimulus is present. For example, a dog raises its ear on hearing doorbell.

1.2.3 Problem Solving

Problem solving is the process in which one perceives and tries to arrive at a desired solution from a present situation by taking some path, which is blocked by known or unknown hurdles. It also includes decision making, which is the process of selecting the best suitable alternative out of multiple alternatives to reach the desired goal available.

1.2.4 Perception

Perception is the process of acquiring, interpreting, selecting, and organizing sensory information.

In humans, perception is aided by sensory organs. In the domain of AI, perception mechanism puts the data acquired by the sensors together in a meaningful manner.

1.3 Difference Between Human and Machine Intelligence

Human perceive by patterns whereas the machines perceive by set of rules and data. Humans store and recall information by patterns, while machines do it by searching algorithms. Human can figure out the complete object even if some part of it is missing or distorted; whereas the machines cannot do it correctly.

2. Artificial Intelligence Basics

Artificial intelligence (AI) is the ability of a machine or computer system to copy human intelligence processes, learn from experiences, adapt to new information, and perform human-like activities. Specific applications of AI include expert systems, natural language processing (NLP), speech recognition and machine vision.

AI is a wide field of study - incorporating various technologies, methods, and theories - that focuses on combining large amounts of data with defined rules and fast, repetitive processing. This enables the software to enhance and improve its ability to complete tasks by recognizing patterns and features in the data sets.

Self-driving cars and computers that play chess are two examples of machines with artificial intelligence. In addition, a variety of industries have begun using AI to improve their work processes — such as healthcare, manufacturing, banking, and retail. Furthermore, AI is finding multiple beneficial uses in cybersecurity.

AI programming focuses on three cognitive skills: learning, reasoning, and self-correction.

- Learning processes. This aspect focuses on acquiring data and creating rules for how to turn the data into actionable information. The rules, which are called algorithms, provide computing devices with step-by-step instructions for how to complete a specific task.
- Reasoning processes. This aspect focuses on choosing the right algorithm to reach a desired outcome.
- Self-correction processes. This aspect is designed to continually fine-tune algorithms and ensure they provide the most accurate results possible.

3. Advantages and Disadvantages of Artificial Intelligence

Artificial neural networks (ANN) and deep learning technologies are quickly evolving, primarily because AI processes large amounts of data much faster and makes predictions more accurately than human. While the huge volume of data that's being created on a daily basis would bury a human researcher, AI applications that use machine learning can take that data and quickly turn it into actionable information.

The primary disadvantage of using AI is that it is expensive to process the large amounts of data that AI programming requires. Other disadvantages include its potential to increase unemployment by replacing jobs previously held by humans; its lack of creativity because machines can only do what they're taught or told; and its inability to completely replicate humans.

AI can be categorized as either weak or strong. Weak AI, also known as narrow AI, is an AI system that is designed and trained to complete a specific task. Industrial robots and virtual personal assistants, such as Apple's Siri, use weak AI.

Strong AI, also known as artificial general intelligence (AGI), describes programming that can replicate human cognitive abilities. When presented with an unfamiliar task, a strong AI system can use fuzzy logic to apply knowledge from one domain to another and find a solution autonomously. In theory, a strong AI program should be able to pass a Turing test.

4. Basic Components of Artificial Intelligence

The five major components that make artificial intelligence a successful one are as follows.

4.1 Discovering

It is the basic ability of an intelligent system to explore the data from available resources without any human intervention. Then it is processed by the ETL algorithm to explore the large database and automatically finds the relationship between the content and the needed solution to the problem. This not only solves a complex issue but also identify the emergency phenomena.

4.2 Predicting

This approach is designed to identify future happenings by classification, ranking, and regression. The algorithm used here is random forest, linear learners and gradient boosting. Rarely prediction goes wrong in some numerical values when there is bias.

4.3 Justifying

Application needs human intervention to give a more recognizable and believable result. So it needs to understand and justify what is wrong and right and then gives human a correct solution to handle the situation. Similar to the automation industry it needs to have nuts and bolts understanding of machine to know why it is repaired and what needs to be done further.

4.4 Acting

Intelligent application needs to be active and live in the company to discover, predict and justify.

4.5 Learning

The intelligent system has the habit of learning and updating itself day by day to compete in the world's needs.

5. Applications of Artificial Intelligence

Below are the various applications of artificial intelligence.

- AI is used in the finance industry where personal data is collected which can be later used to provide financial advice.
- AI is used in the field of education, where the grading system can be automated, the performance of the students can be assessed and the learning process can be improved.
- In the field of healthcare, AI is used to perform a better diagnosis where the technologies are used to understand the natural language and respond to the questions asked. Also, computer programs like chatbots are used to assist customers in scheduling appointments and ease of billing process, etc.
- AI is used in business to automate the repetitive tasks performed by humans with the help of robotic process automation. To increase customer satisfaction, machine learning algorithms are integrated with analytics to gather information which helps in understanding customer needs.
- AI is used in smart home devices, security and surveillance, navigation and travel, music and media streaming and video games, etc.

6. Conclusion

AI is impacting our lives on a great scale. Organizations are also taking steps towards adapting to the AI technology, which can give them new ways of performing the tasks as well as understanding the data pattern to have maximum productivity.

New Words

intelligence	[in'telidʒəns]	n.智能，智力
calculate	['kælkjʊleit]	v.计算；估计
reason	['ri:zn]	v.推理
perceive	[pə'si:v]	v.意识到；察觉，发觉；理解
retrieve	[ri'tri:v]	vt.取回，恢复；检索
comprehend	[ˌkɔmpri'hend]	vt.理解，领会
judgement	['dʒʌdʒmənt]	n.判断，看法，意见
observation	[ˌɔbzə'veiʃn]	n.观察；观察力
statement	['steitmənt]	n.陈述；声明
examine	[ig'zæmin]	v.检查；调查

possibility	[ˌpɔsə'biləti]	n.可能，可能性
logical	['lɔdʒikl]	adj.逻辑（上）的；符合逻辑的
premises	['premisiz]	n.前提
practise	['præktis]	v.练习，实习
auditory	['ɔ:ditəri]	adj.听觉的，听觉器官的
episodic	[ˌepi'sɔdik]	adj.由片段组成的
sequence	['si:kwəns]	n.一系列；一连串；顺序
witness	['witnis]	vt.出席或知道
mimic	['mimik]	vt.模仿，摹拟
		adj.模仿的，摹拟的
perceptual	[pə'septʃuəl]	adj.知觉的，有知觉的
stimuli	['stimjulai]	n.刺激，刺激物；促进因素
absolute	['æbsəlu:t]	adj.绝对的，完全的
spatial	['speiʃl]	adj.空间的
roadmap	['rəudmæp]	n.路标，路线图
hurdle	['hə:dl]	n.障碍，困难
alternative	[ɔ:l'tə:nətiv]	adj.替代的；备选的；其他的
		n.可供选择的事物
perception	[pə'sepʃn]	n.感知，知觉；觉察（力）
sensory	['sensəri]	adj.感觉的，感受的，感官的
sensor	['sensə]	n.传感器
meaningful	['mi:niŋfl]	adj.有意义的；意味深长的
algorithm	['ælgəriðəm]	n.算法
copy	['kɔpi]	v.复制
process	['prəuses]	n.过程
		vt.加工；处理
perform	[pə'fɔ:m]	vt.工作；做；进行；完成；
		vi.运行，表现
activity	[æk'tivəti]	n.行动，活动
application	[ˌæpli'keiʃn]	n.适用，应用，运用
recognition	[ˌrekəg'niʃn]	n.认识，识别
vision	['viʒn]	n.视力，视觉；想象；幻象
method	['meθəd]	n.方法
rule	[ru:l]	n.规则

repetitive	[ri'petətiv]	adj.重复的
pattern	['pætn]	n.模式；图案
		vt.模仿
retail	['ri:teil]	n.零售
		vt.零售；零卖
multiple	['mʌltipl]	adj.多重的；多个的；多功能的
cybersecurity	['saibəsi'kjuərəti]	adj.计算机安全，网络安全
programming	['prəugræmiŋ]	n.编程
cognitive	['kɔgnitiv]	adj.认知的，认识的
aspect	['æspekt]	n.方面
acquire	[ə'kwaiə]	vt.获得，取得
actionable	['ækʃənəbl]	adj.可行动性；可执行的
device	[di'vais]	n.装置，设备
instruction	[in'strʌkʃn]	n.指令
task	[tɑ:sk]	n.工作，任务；作业
continually	[kən'tinjuəli]	adv.连续地，不停地，持续地
fine-tune	['fain'tju:n]	vt.调整，对……进行微调
evolve	[i'vɔlv]	v.发展；进化
prediction	[pri'dikʃn]	n.预测，预报；预言
accurately	['ækjurətli]	adv.准确地，精确地
huge	[hju:dʒ]	adj.巨大的，庞大的，极大的
expensive	[ik'spensiv]	adj.昂贵的，花钱多的
unemployment	[ˌʌnim'plɔimənt]	n.失业；失业率
creativity	[ˌkri:ei'tivəti]	n.创造性，创造力
robot	['rəubɔt]	n.机器人
virtual	['və:tʃuəl]	adj.（计算机）虚拟的
describe	[di'skraib]	vt.描写，形容；叙述
unfamiliar	[ˌʌnfə'miliə]	adj.不熟悉的；不常见的
solution	[sə'lu:ʃn]	n.解决方案
autonomously	[ɔ:'tɔnəməsli]	adv.自主地
component	[kəm'pəunənt]	n.成分；零件；要素
		adj.组成的；构成的
explore	[ik'splɔ:]	v.探索，探究
intervention	[ˌintə'venʃn]	n.介入，干涉，干预

phenomena	[fi'nɔminə]	n.现象
recognizable	['rekəgnaizəbl]	adj.可识别的；可认识的
believable	[bi'li:vəbl]	adj.可信的
justify	['dʒʌstifai]	vt.对……做出解释；证明……有理
habit	['hæbit]	n.习惯，习性
collect	[kə'lekt]	vt.收集；聚积
automate	['ɔ:təmeit]	v.（使）自动化，使自动操作
schedule	['ʃedju:l]	n.日程安排；工作计划
satisfaction	[ˌsætis'fækʃn]	n.满足，满意
surveillance	[sə'veiləns]	n.监督，监视
navigation	[ˌnævi'geiʃn]	n.导航
maximum	['mæksiməm]	adj.最大值的，最大量的

Phrases

make a decision	决策
inductive reasoning	归纳推理
deductive reasoning	演绎推理
be categorized as	被分类为……
sensory organ	感觉器官
figure out	想出；弄明白；解决
adapt to	使适应于，能应付……，变得习惯于……
expert system	专家系统
speech recognition	语音识别
machine vision	机器视觉
focus on	聚焦于，专注于
data set	数据集合
self-driving car	自动驾驶汽车
a variety of	多种的，种种的
deep learning	深度学习
lack of	缺乏
weak AI	弱人工智能
narrow AI	窄人工智能

personal assistant	个人助理
strong AI	强人工智能
fuzzy logic	模糊逻辑
be able to	能，会；能够
Turing test	图灵测试
random forest	随机森林
linear learner	线性学习器
gradient boosting	梯度提升
respond to	对……做出反应，响应
integrate with	（使）与……整合

Abbreviations

AI (Artificial Intelligence)	人工智能
NLP (Natural Language Processing)	自然语言处理
ANN (Artificial Neural Networks)	人工神经网络
AGI (Artificial General Intelligence)	通用人工智能
ETL (Extract, Transform, Load)	数据抽取、转换、装载

Text A 参考译文
人工智能概述

1. 智能

1.1 什么是智能

智能是系统计算、推理、感知关系和类比、从经验中学习、在记忆中存储和检索信息、解决问题、理解复杂思想、分类、概括和适应新情况的能力。

1.2 智能由什么组成

1.2.1 推理

推理是一系列过程，为我们的判断、决策和预测提供基础。推理大致有以下两种类型（请参阅表 1-1）：

表 1-1　　　　　　　　　　　　　　　　　　推理类型

归纳推理	演绎推理
进行特定的观察以做出广泛的一般性陈述	从一般性陈述开始，研究得出具体、合乎逻辑的结论的可能性
即使陈述中的所有前提都是正确的，归纳推理的结论也可能是错误的	如果在一般情况下某件事对一类事物是正确的，那么对于该类的所有成分来说也是正确的

1.2.2　学习

学习是通过学习、练习、听从指导或经历某些事情来获取知识或技能的活动。学习可以提高对研究主题的认识。

人、某些动物和支持 AI 的系统具有学习的能力。学习可以分为：

（1）听觉学习：是通过听来学习。例如，学生听录制好的音频讲座。

（2）情节学习：通过记住一个人目睹或经历过的事件的进展来学习。这是线性且有序的。

（3）运动学习：通过精确的肌肉运动来学习。例如，拾取物体、书写等。

（4）观察学习：通过观察和模仿他人来学习。例如，孩子试图通过模仿父母来学习。

（5）感知学习：即识别一个人以前经历过的刺激。例如，识别和分类对象和情况。

（6）关系学习：它涉及学会根据关系属性而非绝对属性来区分各种刺激。

（7）空间学习：通过视觉刺激（例如图像、颜色、地图等）进行学习。例如，一个人可以在实际走这条路之前在脑海中先创建一个路线图。

（8）刺激反应学习：这是学会在存在某种刺激时执行特定行为的过程。例如，一条狗在听到门铃时竖起耳朵。

1.2.3　解决问题

解决问题是一个过程。在该过程中，人们会通过采取某种方法来感知并尝试从当前情境中获得所需的解决方案，该过程被已知或未知的障碍所阻止。它还包括决策，即从多个备选方案中选择最合适的方案以实现预期目标的过程。

1.2.4　感知

感知是获取、解释、选择和组织感官信息的过程。

对人来说，感知需要感觉器官的协助。在 AI 领域，感知机制将传感器获取的数据以有意义的方式组合在一起。

1.3　人与机器智能的区别

人类通过模式来感知，而机器通过一组规则和数据来感知。人类通过模式存储和调用信息，而机器通过搜索算法来实现。即使部分物体丢失或变形，人类也可以复原出完整的物体。而机器却无法正确做到。

2. 人工智能的基础

人工智能（AI）是机器或计算机系统复制人类智力过程、从经验中学习、适应新信息并进行类似人类活动的能力。AI 的特定应用包括专家系统、自然语言处理（NLP）、语音识别和机器视觉。

人工智能是一个广泛的研究领域——融合了各种技术、方法和理论——致力于将大量数据与已定义的规则结合起来，并进行快速、重复的处理。这使得软件能够通过识别数据集中的模式和功能来增强和提高其完成任务的能力。

自动驾驶汽车和计算机下国际象棋是人工智能机器的两个示例。许多行业（例如医疗保健、制造业、银行业和零售业）已开始使用 AI 来改善其工作流程。另外，人工智能也使网络安全领域获益匪浅。

AI 编程聚焦于 3 种认知技能：学习、推理和自我纠正。

- 学习过程。重点是获取数据并创建将数据转换为可操作信息的规则。这些规则被称为算法，可为计算设备提供逐步完成特定任务的指令。
- 推理过程。着重于选择正确的算法以达成期望的结果。
- 自我纠正过程。旨在不断优化算法，并确保提供最准确的结果。

3. 人工智能的利弊

人工神经网络（ANN）和深度学习技术正在迅速发展，这主要是因为与人类相比，AI 能够处理大量数据并做出更准确的预测。虽然每天创建的海量数据足以淹没研究人员，但使用机器学习的 AI 应用程序可以获取这些数据并将其迅速转变为可操作的信息。

使用 AI 的主要缺点是处理 AI 编程所需的大量数据代价昂贵。其他缺点包括它有可能通过取代人所担任的工作使失业增加；它缺乏创造力，因为机器只能完成人教给它们或告诉它们的事情；它无法完全复制人类。

AI 可以分为强、弱两种。弱 AI（也称为窄 AI）是被设计和训练来完成特定任务的 AI 系统，工业机器人和虚拟个人助理（例如 Apple 的 Siri）使用弱化的 AI。

强 AI（也称为通用人工智能）称作可以复制人类认知能力的编程。当遇到不熟悉的任务时，强 AI 系统可以使用模糊逻辑将知识从一个领域应用于另一个领域，并自动找到解决方案。理论上，一个强 AI 程序应该能够通过图灵测试。

4. 人工智能的基本组成

完备的人工智能具有以下五种要素。

4.1 发现

智能系统的基本功能是无须任何人工干预即可从可用资源中浏览数据。然后，通过 ETL 算法对其进行处理，以探索大型数据库，并自主找到内容与所需解决方案之间的关系。这不仅解决了一个复杂的问题，而且识别了紧急现象。

4.2 预测

该方法旨在通过分类、排名和回归来识别未来事件。这里使用的算法是随机森林、线性学习器和梯度提升。即便在某些数值上有偏差，预测也很少出错。

4.3 判断

应用程序需要人为干预才能给出更好识别和更为可信的结果。因此，它需要了解并证明错误和正确的地方，然后为人类处理这种情况提供正确的解决方案。与自动化行业类似，需要先了解机器的基本组成，才能知道机器为什么需要修理以及需要做些什么。

4.4 行动

应该积极主动地在公司内使用人工智能，以有助于发现、预测和论证。

4.5 学习

智能系统习惯于每天学习和更新以适应外界的需求。

5. 人工智能的应用

以下是人工智能的各种应用。

- 人工智能用于金融行业，收集个人数据，这些数据之后可用于提供金融建议。
- AI 用于教育领域，在该领域中，评分系统可以实现自动化，用以评估学生的表现，并且改善学习过程。
- 在医疗保健领域，人工智能技术用于辅助诊断，用于理解自然语言并回应所提问的问题。同样，诸如聊天机器人之类的计算程序也用于协助客户安排约会并简化计费过程等。
- 在商业中 AI 用于借助机器人流程自动化来自动执行由人类执行的重复性任务。为了提高客户满意度，将机器学习算法与分析功能集成在一起，以收集有助于理解客户需求的信息。
- AI 可用于智能家居设备、安全和监视、导航和旅行、音乐和媒体流以及视频游戏等。

6. 结论

人工智能正在极大地影响着我们的生活。人们也正在采取措施适应 AI 技术，这可以为他们提供执行任务及理解数据模式的新方法，实现最大的生产率。

Text B
Artificial Intelligence Applications in the Real World

1. Marketing

扫码听课文

In the early 2000s, if we searched an online store to find a product without knowing its exact name, it would become a nightmare to find the product. But now when we search for an item on any e-commerce store, we get all possible results related to the item. It's like these search engines read our minds! In a matter of seconds, we get a list of all relevant items. An example of this is finding the right movies on Netflix.

One reason why we're all obsessed with Netflix is because Netflix provides highly accurate predictive technology based on customer's reactions to films. It examines millions of records to suggest shows and films that you might like based on your previous actions and choices of films. As the data set grows, this technology is getting smarter and smarter every day.

2. Banking

AI in banking is growing faster than you thought! A lot of banks have already adopted AI-based systems to provide customer support, detect anomalies and credit card frauds. An example of this is HDFC Bank.

HDFC Bank has developed an AI-based chatbot called EVA (Electronic Virtual Assistant), built by Bengaluru-based Senseforth AI Research.

Since its launch, Eva has addressed over 3 million customer queries, interacted with over half a million unique users, and held over a million conversations. Eva can collect knowledge from thousands of sources and provide simple answers in less than 0.4 seconds.

The use of AI for fraud prevention is not a new concept. In fact, AI solutions can be used to enhance security across a number of business sectors, including retail and finance.

By tracing card usage and endpoint access, security specialists are more effectively preventing fraud. Organizations rely on AI to trace those steps by analyzing the behaviors of transactions.

Companies such as MasterCard and RBS WorldPay have relied on AI and deep learning to detect fraudulent transaction patterns and prevent card fraud for years now. This has saved millions of dollars.

3. Finance

Ventures have been relying on computers and data scientists to determine future patterns in the market. Trading mainly depends on the ability to predict the future accurately.

Machines are great at this because they can crunch a huge amount of data in a short span. Machines can also learn to observe patterns in past data and predict how these patterns might repeat in the future.

In the age of ultra-high-frequency trading, financial organizations are turning to AI to improve their stock trading performance and boost profit.

One such organization is Japan's leading brokerage house, Nomura Securities. The company has been pursuing one goal, i.e. to analyze the insights of experienced stock traders with the help of computers. After years of research, Nomura is set to introduce a new stock trading system.

The new system stores a vast amount of price and trading data in its computer. By tapping into this reservoir of information, it will make assessments, for example, it may determine that current market conditions are similar to the conditions two weeks ago and predict how share prices will be changing a few minutes down the line. This will help to take better trading decisions based on the predicted market prices.

4. Agriculture

AI can help farmers get more from the land while using resources more sustainably.

Organizations are using automation and robotics to help farmers find more efficient ways to protect their crops from weeds.

Blue River Technology has developed a robot called See & Spray which uses computer vision technologies like object detection to monitor and precisely spray weedicide on cotton plants. Precision spraying can help prevent herbicide resistance.

Apart from this, a Berlin-based agricultural tech start-up called PEAT has developed an application called Plantix that identifies potential defects and nutrient deficiencies in the soil through images.

The image recognition App identifies possible defects through images captured by the user's smartphone camera. Users are then provided with soil restoration techniques, tips, and other possible solutions. The company claims that its software can achieve pattern detection with an estimated accuracy of up to 95%.

5. Health Care

When it comes to saving our lives, a lot of organizations and medical care centers are relying on AI. There are many examples of how AI in healthcare has helped patients all over the world.

An organization called Cambio Health Care developed a clinical decision support system for stroke prevention that can give the physician a warning when there's a patient at risk of having a heart stroke.

Another such example is Coala life which is a company that has a digitalized device that can find cardiac diseases.

6. Gaming

Over the past few years, artificial intelligence has become an integral part of the gaming industry. In fact, one of the biggest accomplishments of AI is in the gaming industry.

DeepMind's AI-based AlphaGo software, which is known for defeating Lee Sedol, the world champion in the game of GO, is considered to be one of the most significant accomplishments in the field of AI.

Shortly after the victory, DeepMind created an advanced version of AlphaGo called AlphaGo Zero which defeated the predecessor in an AI-AI face off. Unlike the original AlphaGo, which DeepMind trained over time by using a large amount of data and supervision, the advanced system, AlphaGo Zero taught itself to master the game.

7. Space Exploration

Space expeditions and discoveries always require analyzing vast amounts of data. Artificial intelligence is the best way to handle and process data on this scale. After rigorous research, astronomers use artificial intelligence to sift through years of data obtained by the Kepler telescope in order to identify a distant eight-planet solar system.

8. Autonomous Vehicles

For a long time, self-driving cars have been a buzzword in the AI industry. The development of autonomous vehicles will definitely revolutionize the transport system.

Companies like Waymo conducted several test drives in Phoenix before deploying their first AI-based public ride-hailing service. The AI system collects data from the vehicles radar, cameras, GPS, and cloud services to produce control signals that operate the vehicle.

Advanced deep learning algorithms can accurately predict what objects in the vehicle's vicinity are likely to do. This makes Waymo cars more effective and safer.

Another famous example of an autonomous vehicle is Tesla's self-driving car. Artificial intelligence implements computer vision, image detection, and deep learning to build cars that can automatically detect objects and drive around without human intervention.

9. Chatbots

These days virtual assistants have become a very common technology. Almost every household has a virtual assistant that controls the appliances at home. A few examples include Siri, Cortana, which are gaining popularity because of the user experience they provide.

Amazon's Echo is an example of how artificial intelligence can be used to translate human language into desirable actions. This device uses speech recognition and NLP to perform a wide range of tasks on your command. It can do more than just play your favorite songs. It can be used to control the devices at your house, book cabs, make phone calls, order your favorite food, check the weather conditions and so on.

Another example is the newly released Google's virtual assistant called Google Duplex, which has astonished millions of people. Not only can it respond to calls and book appointments for you, but it also adds a human touch.

The device uses natural language processing and machine learning algorithms to process human language and perform tasks such as manage your schedule, control your smart home, make a reservation and so on.

10. Social Media

Ever since social media has become our identity, we've been generating an immeasurable amount of data through chats, tweets, posts and so on. And wherever there is an abundance of data, AI and machine learning are always involved.

In social media platforms like Facebook, AI is used for face verification wherein machine learning and deep learning concepts are used to detect facial features and tag your friends. Deep learning is used to extract every minute detail from an image by using a bunch of deep neural networks. On the other hand, machine learning algorithms are used to design your feed based on your interests.

New Words

search	[sə:tʃ]	v.搜寻，搜索
nightmare	['naitmeə]	n.噩梦；可怕的事情
e-commerce	[i:'kɔmə:s]	n.电子商务
engine	['endʒin]	n.发动机，引擎

relevant	['reləvənt]	adj.有关的，相关联的
accurate	['ækjʊrət]	adj.精确的，准确的
record	['rekɔːd]	n.记录
detect	[dɪ'tekt]	vt.查明，发现；侦察，侦查
anomaly	[ə'nɔməli]	n.异常，反常
fraud	[frɔːd]	n.欺诈；骗子；伪劣品；冒牌货
conversation	[ˌkɔnvə'seɪʃn]	n.交谈，会话；交往，交际；人机对话
prevention	[prɪ'venʃn]	n.预防；阻止，制止
enhance	[ɪn'hɑːns]	vt.提高，增加，加强
security	[sɪ'kjʊərəti]	n.安全；保证
fraudulent	['frɔːdjulənt]	adj.欺骗的，不诚实的
crunch	[krʌntʃ]	vt.处理
repeat	[rɪ'piːt]	vt.重复
boost	[buːst]	vt.促进，提高，增加
profit	['prɔfɪt]	n.收益，利润
reservoir	['rezəvwɑː]	n.蓄水池，水库
assessment	[ə'sesmənt]	n.评估，评价
determine	[dɪ'tɜːmɪn]	vt.确定；判定
sustainably	[səs'teɪnəbli]	adv.可持续地
crop	['krɔp]	n.作物
weedicide	['wiːdɪsaɪd]	n.除草剂
herbicide	['hɜːbɪsaɪd]	n.除草剂
nutrient	['njuːtrɪənt]	n.养分，养料
deficiency	[dɪ'fɪʃnsi]	n.缺乏，不足；缺点，缺陷
restoration	[ˌrestə'reɪʃn]	n.恢复；复原；整修
estimate	['estɪmət]	n.估计，预测
	['estɪmeɪt]	vt.估计，估算
stroke	[strəʊk]	n.中风
accomplishment	[ə'kʌmplɪʃmənt]	n.成就；完成；技能；履行
predecessor	['priːdɪsesə]	n.前任，前辈，前身
rigorous	['rɪɡərəs]	adj.严密的，缜密的
astronomer	[ə'strɔnəmə]	n.天文学者，天文学家
telescope	['telɪskəʊp]	n.望远镜
buzzword	['bʌzwɜːd]	n.时髦术语，流行行话
autonomous	[ɔː'tɔnəməs]	adj.自治的；有自主权的
definitely	['defɪnɪtli]	adv.明确地；确切地；一定地
revolutionize	[ˌrevə'luːʃənaɪz]	v.彻底改变；完全变革

signal	['signəl]	n.信号
		vt.向……发信号
		vi.发信号
vicinity	[vi'sinəti]	n.附近，邻近
astonish	[ə'stɔniʃ]	vt.使惊讶，使大为吃惊
reservation	[ˌrezə'veiʃn]	n.保留；预订，预约
identity	[ai'dentəti]	n.身份
post	[pəust]	n.帖子
abundance	[ə'bʌndəns]	n.丰富，充裕；大量
verification	[ˌverifi'keiʃn]	n.证明；证实；核实
tag	[tæg]	vt.加标签于
		n.标签
interest	['intrist]	n.兴趣，爱好

Phrases

online store	在线商店
be obsessed with	痴迷于
base on	基于……，建立在……上
AI-based system	基于人工智能的系统
credit card	信用卡
interact with	与……互动，与……相互影响，与……相互配合
rely on	信赖，依赖，依靠
data scientist	数据科学家
depend on	依赖；相信
in a short span	在短时间内
stock trading	股票交易
decision support system	决策支持系统
digitalized device	数字化设备
cardiac disease	心脏病
be known for	因……而众所周知，因……而出名
space expedition	太空探险
transport system	运输系统
smart home	智能家居

social media 社交媒体
a bunch of 一群；一束；一堆

Abbreviations

EVA (Electronic Virtual Assistant)   电子虚拟助理

Text B 参考译文
现实世界中人工智能的应用

1. 营销

在 21 世纪初期，如果我们在不知道确切名称的情况下搜索在线商店来找一个产品，那很难找到该产品。但是现在，当我们在任何电子商务商店中搜索商品时，我们都会获得与该商品相关的所有可能结果。就像这些搜索引擎明白了我们的心思！在短暂的几秒内，我们将获得所有相关项目的列表。例如，我们可以在 Netflix 上找到合适的电影。

我们沉迷于 Netflix 的原因之一是 Netflix 根据客户对电影的反应提供了高度准确的预测技术。它会检查数百万条记录，根据你以前的操作和选择的电影来推荐你可能会喜欢的表演和电影。随着数据集的增长，这项技术每天都变得越来越智能。

2. 银行业务

银行业的 AI 增长速度超出你的想象！许多银行已经采用基于 AI 的系统给客户提供支持，检测异常情况和信用卡欺诈。HDFC 银行就是一个例子。

HDFC 银行开发了一个基于 AI 的聊天机器人，称为 EVA（电子虚拟助手），由位于班加罗尔的 Senseforth AI Research 构建。

自推出以来，Eva 已经处理的客户查询超过 300 万个，与超过 50 万的独特用户进行了互动，并进行了超过 100 万次的对话。Eva 可以从数千个来源中收集知识，并在 0.4 秒内提供简明的答案。

使用 AI 预防欺诈不是一个新概念。实际上，人工智能解决方案可用于增强零售和金融等多个业务部门的安全性。

通过跟踪卡的使用和端点访问，安全专家可以更有效地防止欺诈。组织依靠 AI 通过分析交易行为来追踪这些步骤。

万事达卡（MasterCard）和苏格兰皇家银行（RBS）WorldPay 等公司多年来一直依靠 AI 和深度学习来检测欺诈性交易模式并防止卡欺诈，这节省了数百万美元。

3. 金融

风险投资一直依靠计算机和数据科学家来确定市场的未来模式。交易主要取决于准确预测未来的能力。

计算机之所以出色，是因为它们可以在短时间内处理大量数据。计算机还可以学习观察过去数据中的模式，并预测这些模式在未来可能重复的方式。

在超高频交易时代，金融机构正在转向使用 AI 来改善其股票交易性能并提高利润。

日本领先的经纪公司野村证券就是这样的组织。该公司一直在瞄定一个目标，即借助计算机来分析经验丰富的股票交易员的见解。经过多年的研究，野村证券将推出一种新的股票交易系统。

新系统在其计算机中存储了大量的价格和交易数据，通过该信息库，它将进行评估，例如，它可以确定当前市场状况与两周前的状况相似，并预测股价在几分钟内将如何变化。这将有助于根据预测的市场价格做出更好的交易决策。

4. 农业

人工智能可以帮助农民从土地上获得更多收益，同时更有效地持续使用资源。

组织正在使用自动化和机器人技术来帮助农民找到更有效的方法，以保护农作物免受杂草侵害。

蓝河技术公司已经开发出一种名为 See & Spray 的机器人，该机器人使用诸如对象检测之类的计算机视觉技术来监控并精确地把除草剂喷洒到棉株上。精确喷雾有助于防止除草剂抗性。

除此之外，位于柏林的农业科技初创公司 PEAT 开发了一个名为 Plantix 的应用程序，该应用程序可以通过图像识别出土壤中潜在的缺陷和营养缺乏症。

图像识别应用可通过用户智能手机的相机捕获的图像来识别可能的缺陷，然后为用户提供土壤修复技术、技巧和其他可能的解决方案。该公司声称其软件可以实现模式检测，估计精度高达 95%。

5. 医疗保健

在挽救生命方面，许多组织和医疗中心都依赖于 AI，有很多例证都说明了人工智能在医疗保健中是如何帮助全世界的患者的。

一家名为 Cambio 医疗的组织开发了一种用于预防中风的临床决策支持系统，该系统可以在人们具有患中风的风险时向医生发出警告。

另一个类似的例子是 Coala life，这是一家拥有可以检测心脏病的数字化设备的公司。

6. 游戏

在过去的几年中，人工智能已成为游戏行业不可或缺的一部分。实际上，人工智能的最大成就之一就是在游戏行业。

DeepMind 基于 AI 的 AlphaGo 软件以击败围棋世界冠军李世石而闻名，被认为是 AI 领域最重要的成就之一。

获胜后不久，DeepMind 创建了一个 AlphaGo 高级版本，被称为 AlphaGo Zero。该版本在 AI 对 AI 的对抗中击败了 AlphaGo。DeepMind 通过使用海量数据和监督一直对原来的 AlphaGo 进行着训练，而先进的 AlphaGo Zero 则与之不同，它能够自学掌握游戏的知识。

7. 太空探索

太空探险和发现总是需要分析大量数据，人工智能是处理和加工这种规模数据的最佳方法。经过严格的研究，天文学家使用人工智能筛选通过开普勒望远镜获得的多年数据，以识别遥远的有八大行星的太阳系。

8. 自动驾驶汽车

在很长的时间内，自动驾驶汽车一直是 AI 行业的流行语。自动驾驶汽车的发展必将彻底改变交通运输系统。

诸如 Waymo 之类的公司在部署其首个基于 AI 的公共乘车服务之前，已在凤凰城进行了几次试驾。AI 系统从车辆雷达、摄像机、GPS 和云服务收集数据，以产生操作车辆的控制信号。

先进的深度学习算法可以准确预测车辆附近可能出现的物体的行动。这使 Waymo 汽车更有效、更安全。

自动驾驶汽车的另一个著名例子是特斯拉的自动驾驶汽车。人工智能实现计算机视觉、图像检测和深度学习，以制造能够自动检测物体并在无人干预的情况下四处行驶的汽车。

9. 聊天机器人

如今，虚拟助手已成为一种非常普遍的技术。几乎每个家庭都有一个虚拟助手来控制家里的设备。一些例子包括 Siri 和 Cortana，它们由于提供了很好的用户体验而越来越受欢迎。

亚马逊的 Echo 也是一个例子，表明了人工智能怎样将人类语言转化为令人满意的动作。该设备使用语音识别和自然语言处理来按照你的命令执行各种任务，它不仅可以播放你喜欢的歌曲，还可以做更多的事情。它可用于控制房屋中的设备，预订出租车，打电话，订购你喜欢的食物，检查天气状况等。

另一个例子是新发布的 Google 虚拟助手 Google Duplex，让无数人震惊不已。它不仅可以为你接听电话和预定约会，而且增加了人情味。

该设备使用自然语言处理和机器学习算法来处理人类语言并执行诸如管理日程安排、控制智能家居、进行预订等任务。

10. 社交媒体

自从社交媒体成为我们的身份标识以来，我们一直在通过聊天、推文、帖子等方式生成海量数据。哪里有海量数据，哪里就会涉及 AI 和机器学习。

在 Facebook 之类的社交媒体平台中，人工智能用于面部验证，其中之一就是使用机器学习和深度学习概念检测面部特征并标记你的朋友。深度学习用于通过使用一组深度神经网络从图像中提取每一个细节，另外，机器学习算法用于根据你的兴趣来设计推送。

Exercises

[Ex. 1] Answer the following questions according to Text A.

1. What is intelligence?
2. What is intelligence composed of?
3. What is reasoning? What are the two broadly types of reasoning?
4. What is learning?
5. What is artificial intelligence (AI)? What do specific applications of AI include?
6. What does AI programming focus on?
7. What is the primary disadvantage of using AI?
8. What is weak AI? What does strong AI describe?
9. What are the five major components that make artificial intelligence a successful one?
10. What are the fields AI is used in according to the passage?

[Ex. 2] Fill in the following blanks according to Text B.

1. One reason why we're all obsessed with Netflix is because Netflix provides _____ based on customer's reactions to films. It examines millions of records to suggest shows and films that you might like based on _____ and _____.
2. HDFC Bank has developed an AI-based chatbot called _____, built by Bengaluru-based Senseforth AI Research. Eva can _____ from thousands of sources and provide simple answers in _____.
3. Ventures have been relying on computers and data scientists to _____ in the

market. Trading mainly depends on the ability to _____.

4. AI can help farmers _____ while using resources more sustainably. Organizations are using _____ to help farmers find more efficient ways to _____.

5. An organization called Cambio Health Care developed _____ for stroke prevention that can _____ when there's a patient at risk of _____.

6. Over the past few years, artificial intelligence has become an integral part of _____. DeepMind's AI-based AlphaGo software, which is known for defeating Lee Sedol, _____, in the game of GO, is considered to be one of _____ in the field of AI.

7. Artificial intelligence implements _____, _____ and _____ to build cars that can automatically detect objects and drive around _____.

8. Amazon's Echo is an example of how artificial intelligence can be used to translate _____ into _____. This device uses _____ and _____ to perform a wide range of tasks on your command.

9. Google Duplex has astonished millions of people. Not only can it _____ and _____, but it also adds a human touch. The device uses _____ and _____ to process human language and perform tasks such as _____, control your smart home, make a reservation and so on.

10. In social media platforms like Facebook, AI is used for _____ wherein machine learning and deep learning concepts are used to _____ and _____. Deep learning is used to extract every minute detail from an image by _____.

[Ex. 3] Translate the following terms or phrases from English into Chinese and vice versa.

1. making decision   1. _____
2. deductive reasoning   2. _____
3. deep learning   3. _____
4. expert system   4. _____
5. machine vision   5. _____
6. v.（使）自动化，使自动操作   6. _____
7. n.算法   7. _____
8. n.模式；图案 vt.模仿   8. _____
9. n.感知，知觉；觉察（力）   9. _____
10. adj.（计算机）虚拟的   10. _____

[Ex. 4] Translate the following passages into Chinese.

Machine Vision

Machine vision is the ability of a computer to see; and it employs one or more video cameras, analog-to-digital conversion (ADC) and digital signal processing (DSP). The resulting data goes to a computer or robot controller. Machine vision is similar in complexity to voice recognition.

Two important specifications in any vision system are the sensitivity and the resolution. Sensitivity is the ability of a machine to see in dim light, or to detect weak impulses at invisible wavelengths. Resolution is the extent to which a machine can differentiate between objects. In general, the better the resolution, the more confined the field of vision. Sensitivity and resolution are interdependent. All other factors held constant, increasing the sensitivity reduces the resolution, and improving the resolution reduces the sensitivity.

Human eyes are sensitive to electromagnetic wavelengths ranging from 390 to 770 nanometers (nm). Video cameras can be sensitive to a range of wavelengths much wider than this. Some machine-vision systems function at infrared (IR), ultraviolet (UV), or X-ray wavelengths.

Binocular (stereo) machine vision requires a computer with an advanced processor. In addition, high-resolution cameras, a large amount of random access memory (RAM), and artificial intelligence (AI) programming are required for depth perception.

Machine vision is used in various industrial and medical applications. Examples include: electronic component analysis, signature identification, optical character recognition, handwriting recognition, object recognition, pattern recognition, materials inspection, currency inspection, medical image analysis.

The term machine vision is often associated with industrial applications of a computer's ability to see, while the term computer vision is often used to describe any type of technology in which a computer is tasked with digitizing an image, processing the data it contains and taking some kind of action.

[Ex. 5] Fill in the blanks with the words given below.

| increase | sensitivity | conversion | neural | repetitive |
| check | algorithm | recognize | machine | human |

Top 4 Techniques of Artificial Intelligence

1. Machine Learning

It is one of the applications of AI where machines are not explicitly programmed to perform certain tasks rather they learn and improve from experience automatically. Deep Learning is a subset of ___1___ learning based on artificial neural networks for predictive analysis. There are various machine learning algorithms such as unsupervised learning, supervised learning, and reinforcement learning. In unsupervised learning, the ___2___ does not use classified information to act on it without any guidance. In supervised learning, it deduces a function from the training data which consists of a set of an input object and the desired output. Reinforcement learning is used by machines to take suitable actions to ___3___ the reward to find the best possibility which should be taken into account.

2. NLP (Natural Language Processing)

It is the interactions between computers and human language where the computers are programmed to process natural languages. Machine learning is a reliable technology for Natural Language Processing to obtain meaning from ___4___ languages. In NLP, the audio of a human talk is captured by the machine. Then the audio of text conversation occurs and then the text is processed where the data is converted into audio. Then the machine uses the audio to respond to humans. Applications of natural language processing can be found in IVR (Interactive Voice Response) applications used in call centers, language translation applications like Google Translate and word processors such as Microsoft Word to ___5___ the accuracy of grammar in text. However, the nature of human languages makes the Natural Language Processing difficult because of the rules which are involved in the passing of information using natural language, and they are not easy for the computers to understand. So NLP uses algorithms to ___6___ and abstract the rules of the natural languages where the unstructured data from the human languages can be converted to a format that is understood by the computer.

3. Automation and Robotics

The purpose of Automation is to get the monotonous and ___7___ tasks done by machines which also improve productivity and in receiving cost-effective and more efficient results. Many organizations use machine learning, ___8___ networks, and graphs in automation. Such automation can prevent fraud issues in financial transactions online by using CAPTCHA technology. Robotic process automation is programmed to perform high volume repetitive tasks which can adapt to the change in different circumstances.

4. Machine Vision

Machines can capture visual information and then analyze it. Here cameras are used to capture the visual information, the analog to digital ___9___ is used to convert the image to digital data and digital signal processing is employed to process the data. Then the resulting data is fed to a computer. In machine vision, two vital aspects are ___10___, which is the ability of the machine to perceive impulses that are weak and resolution, the range to which the machine can distinguish the objects. The usage of machine vision can be found in signature identification, pattern recognition, and medical image analysis, etc.

Unit 2
Types, Pros and Cons of Artificial Intelligence

Text A
Types of Artificial Intelligence

扫码听课文

The main aim of artificial intelligence is to enable machines to perform a human-like function. Thus the primary way of classifying AI is based on how well it is able to replicate human-like actions. AI can, by and large, be classified based on two types, both of which are based on its ability to replicate the human brain. One type of classification which is "Based on Functionality" classify AI on the basis of their likeness to the human mind and their ability to think and feel like humans. The second way of classification is more prominent in the tech industry which is "Based on Capabilities" of AI in relation to human intelligence.

1. Based on Functionality

1.1 Reactive Machine

This is the most basic and oldest type of artificial intelligence. It replicates a human's ability to react to different kinds of stimuli. This type of AI has no memory power so it lacks the capability to use previously gained information/experience to obtain better results. Therefore this kind of AI doesn't have the ability to train itself like the one we come across nowadays.

Example: Deep Blue, IBM's chess-playing supercomputer, is the perfect example of this kind of machines. It is famous for defeating international grandmaster Garry Kasparov in the late 1990s. Deep Blue can identify different pieces in the chessboard and how each moves. It can identify all the possible legal moves for itself and its opponents. Based on the option, it selects the best possible move. However, it doesn't have the ability to learn from its past moves as this machine doesn't have any memory of its own.

1.2 Limited Theory

This type of AI has memory capabilities so it can use past information/experience to make better future decisions. Most of the common applications existing around us fall under this category. These AI applications can be trained by a large volume of training data which they store in their memory in the form of a reference model.

Example: Limited memory technology is used in many self-driving cars. They store data like GPS location, speed of nearby cars, size /nature of obstructions among a hundred other kinds of data to drive just like a human does.

1.3 Theory of Mind

Theory of mind is the next level of AI, which is very limited, and it doesn't exist in our daily lives. It is mostly in the "Work in Progress" stage and is usually confined to research labs. This kind of AI once developed will have a very deep understating of human minds ranging from their needs, likes, emotions, thought process, etc. Based on its understanding of human minds and their whims AI will be able to alter its own response.

Example: Researcher Winston in his research showed a prototype of a robot which can walk down the small corridor with other robots coming from the opposite direction, the AI can foresee other robots movements and can turn right, left or any other way so as to avoid a possible collision with the incoming robots. As per Wilson, this robot determines its action based on its "common sense" of how other robots will move.

1.4 Self-Aware Artificial Intelligence

This is the final stage of AI. Its current existence is only hypothetical and can be found only in science fiction movies. This kind of AI can not only understand and evoke human emotions but can also have emotions of its own. This kind of AI is decades if not centuries away from materializing. This is because once it is self-aware, the AI can get into self-preservation mode and it might consider humanity as a potential threat and may directly or indirectly pursue endeavor to end humanity.

2. Based on Capabilities

2.1　Artificial Narrow Intelligence (ANI)

All the existing AI applications which we see around us fall under this category. ANI includes an AI system that can perform narrowly defined specific tasks just like humans. However, these machines cannot perform tasks for which it was not programmed beforehand, so they fail at performing unprecedented task. Based on the classification mentioned above, this system is the combination of all reactive and limited memory AI. AI algorithms which we use in today's world for performing most complex prediction modelling fall under this category of AI.

2.2　Artificial General Intelligence (AGI)

AGI has the capability to train, learn, understand, and perform functions just like a normal human does. These systems will have multifunctional capabilities cutting across different domains. They will be more agile and will react and improvise just like humans while facing unprecedented scenarios. There is no real-world example of this kind of AI but good amount of progress has been made in this field.

2.3　Artificial Super Intelligence (ASI)

Artificial super intelligence will be the top-most point of AI development. ASI will be the most potent form of intelligence to ever exist on this planet. It will be able to perform all the tasks better than humans because of their inordinately superior data processing, memory, and decisionmaking ability. Some of the researchers fear that the advent of ASI will ultimately result in "Technological Singularity". It is a hypothetical situation in which the growth in technology will reach an uncontrollable stage which will result in an unimaginable change in human civilization.

At present, it is very hard to foresee how our future will look like when a more dexterous form of AI materializes. However, one can say with great certainty that we are still a long distance apart from reaching that stage as we are just in the very nascent stage on the development of advanced AI. For the proponents of AI, we can say that we are just scratching the surface to unearth the true potential of AI and for the AI skeptics it is too soon to get chills about technological singularity.

New Words

replicate　　　　　　['replikeit]　　　　　　*vt.*复制

	['replikit]	adj.复制的
functionality	[ˌfʌŋkʃə'næləti]	n.功能，功能性
likeness	['laiknis]	n.相像，相似
prominent	['prɔminənt]	adj.突出的，杰出的
experience	[ik'spiəriəns]	n.经验，体验；经历
supercomputer	[ˌsuːpəkəm'pjuːtə]	n.超级计算机
defeat	[di'fiːt]	vt.击败，战胜
international	[ˌintə'næʃnəl]	adj.国际的
grandmaster	['ɡrændˌmɑːstə]	n.特级大师
chessboard	['tʃesbɔːd]	n.棋盘
opponent	[ə'pəunənt]	n.对手，敌手
category	['kætiɡəri]	n.类型，种类，类别
reference	['refrəns]	n.参考；提及，涉及
self-driving	[ˌself'draiviŋ]	adj.自驾的
store	[stɔːr]	v.贮存，存储
prototype	['prəutətaip]	n.原型，雏形，蓝本
opposite	['ɔpəzit]	adj.相对的；对面的
collision	[kə'liʒn]	n.碰撞；冲突
self-aware	[ˌselfə'weə]	adj.自我感知的
evoke	[i'vəuk]	vt.产生，引起；唤起
materialize	[mə'tiəriəlaiz]	vi.具体化；实质化
self-preservation	['selfˌprezə'veiʃn]	n.自保存
potential	[pə'tenʃl]	adj.潜在的，有可能的
pursue	[pə'sjuː]	v.继续；追求
endeavor	[in'devə]	v.尝试，试图；尽力，竭力
		n.努力，尽力
humanity	[hjuː'mænəti]	n.人类；人性；人道
unprecedented	[ʌn'presidentid]	adj.前所未有的，无前例的，空前的
combination	[ˌkɔmbi'neiʃn]	n.组合，结合；联合体
multifunctional	[ˌmʌlti'fʌŋkʃənl]	adj.多功能的
potent	['pəutnt]	adj.有效的，强有力的
inordinate	[in'ɔːdinət]	adj.过度的；过分的；超乎预料的
hypothetical	[ˌhaipə'θetikl]	adj.假设的，假定的；有前提的

uncontrollable	[ˌʌnkən'trəʊləbl]	adj.无法控制的
unimaginable	[ˌʌni'mædʒinəbl]	adj.难以想象的，想不到的
foresee	[fɔː'siː]	vt.预知，预见
dexterous	['dekstrəs]	adj.灵巧的，敏捷的
nascent	['næsnt]	adj.初期的；初生的；开始形成的
proponent	[prə'pəʊnənt]	n.支持者，拥护者，提倡者
skeptic	['skeptik]	n.怀疑者，怀疑论者
singularity	[ˌsɪŋɡjʊ'lærəti]	n.奇点；奇特，奇怪；异常
chill	[tʃɪl]	n.寒冷，恐惧

Phrases

reactive machine	反应机
react to ...	对……做出反应
be famous for ...	因……而著名
be trained by ...	用……训练
reference model	参考模型
theory of mind	心智理论
work in progress	进行中，在路上
be confined to ...	限于……之内
common sense	常识
prediction modelling	预测模型
cut across	跨越
technological singularity	技术奇点

Abbreviations

GPS (Global Position System)	全球定位系统
ANI (Artificial Narrow Intelligence)	窄人工智能
AGI (Artificial General Intelligence)	通用人工智能
ASI (Artificial Super Intelligence)	超人工智能

Text A 参考译文
人工智能的类型

人工智能的主要目的是使机器能够执行类似于人类的功能，因此，对人工智能进行分类的主要方法是基于它能够复制类人行为的能力。大体上，人工智能基于其复制人脑的能力分为两类。第一种分类是"基于功能"，根据人工智能与人类思想的相似程度以及它们思考和感觉类似人类的能力对人工智能进行分类。第二种分类在科技行业中更为突出，这是"基于能力"的相对于人类智能的人工智能。

1. 基于功能

1.1 反应机

它是最基础、最古老的人工智能类型，它复制了人类对不同种类的刺激做出反应的能力。这种类型的人工智能没有记忆能力，因此不具备使用以前已有的信息/经验来获得更好结果的能力，因此，这种类型的人工智能无法像我们今天看到的那样能够进行自我训练。

示例：IBM 的国际象棋超级计算机"深蓝"是这类机器的完美示例，它以在 20 世纪后期击败国际大师加里·卡斯帕罗夫而闻名。深蓝可以识别棋盘中的不同棋子以及棋子的移动方式，它可以确定自己及其对手所有符合规则的可能移动，基于该选项，它将选择最佳移动。但是，它没有自己的记忆，因此无法从过去的移动中学习。

1.2 有限理论

这种类型的人工智能具有存储功能，因此它们可以利用过去的信息/经验来对未来做出更好的决策。我们周围存在的大多数常见应用程序都属于此类。这些人工智能应用程序可以通过大量的训练数据进行训练，这些数据以参考模型的形式存储在内存中。

示例：有限内存技术被用于许多自动驾驶汽车中。它们存储的数据包括 GPS 位置、附近汽车的速度、障碍物的大小/性质以及其他数百种数据，从而就可以像人类那样驾驶。

1.3 心智理论

心智理论是人工智能的下一个层次，它的应用很有限，在我们的日常生活中不存在。这类人工智能大多处于"进行中"阶段，通常仅限于研究实验室。这种类型的人工智能一旦开发出来，它就会对人类的思想有很深的理解，包括人的需求、喜好、情感、思维过程等。基于对人类思想和情绪的理解，人工智能将能够改变自己的反应。

示例：研究人员 Winston 在研究中展示了一种机器人的原型，该机器人可以与来自相反方向的其他机器人一起沿着小走廊走下去，人工智能可以预见其他机器人的运动，并且可以向右、向左或以任何其他方式转弯以避免与对面的机器人发生碰撞。根据 Wilson 的说法，该机器人根据其对其他机器人运动方式的"常识"来确定其动作。

1.4 自我感知的人工智能

这是人工智能的最后阶段，当前只是假设，只存在于科幻电影中。这类人工智能不仅可以理解和唤起人类的情感，还可以拥有自己的情感。这种人工智能成为现实还需要几十年甚至上百年的时间。这是因为，一旦具有自我意识，人工智能就可以进入自我保存模式，它可能把人类当作潜在威胁，并可能直接或间接地致力于终结人类。

2. 基于能力

2.1 窄人工智能（ANI）

我们周围看到的所有现有人工智能应用程序都属于此类。窄人工智能包括一个人工智能系统，该系统可以像人类一样执行明确定义的特定任务。但是，这些机器无法执行未事先编程的任务，因此无法执行新的任务。基于上述分类，该系统是所有反应式和有限内存人工智能的结合。我们当今用于执行最复杂的预测建模的人工智能算法就属于此类人工智能。

2.2 通用人工智能（AGI）

通用人工智能像普通人一样具有训练、学习、理解和执行功能的能力。这些系统将具有跨不同领域的多功能能力，它们将更加敏捷，在面对前所未有的场景时将像人类一样做出反应和即兴发挥。目前尚无此类人工智能的实际示例，但该领域已取得了很大进展。

2.3 超人工智能（ASI）

超人工智能将成为人工智能发展的最高点。超人工智能将是地球上有史以来最强大的智能形式，由于其卓越的数据处理、存储和决策能力，它将能够比人类更出色地执行所有任务。一些研究人员担心超人工智能的出现最终会导致"技术奇点"，这是一种假定的情况，即技术的发展将达到不可控制的阶段，并最终导致人类文明发生无法想象的变化。

目前，很难预见当更灵巧的人工智能形式出现时，我们的未来会是什么样。但是，可以肯定地说，我们正处于距此甚远的阶段，因为我们正处于高级人工智能发展的起步阶段。对于支持人工智能的人来说，可以说我们只是在表层探索以发掘人工智能的真正潜力，对于人工智能怀疑论者来说，现在对技术奇点感到畏惧还为时过早。

Text B
Pros and Cons of Artificial Intelligence

1. Advantages of Artificial Intelligence

1.1 To "Err" Is Human, So Why Not Use Artificial Intelligence

扫码听课文

Machines make decision based on previous data records. With algorithms, the chances of errors are reduced. This is an achievement, as solving complex problems require difficult calculation that should be done without any error. Business organizations use digital assistants to interact with their users, this helps them to save an ample amount of time. The demand for user's businesses is fulfilled and thus they don't have to wait. They are programmed to give the best possible assistance to a user.

1.2 Artificial Intelligence Doesn't Get Tired and Wear out Easily

Artificial intelligence and the science of robotics are used in mining and other fuel exploration processes. These complex machines help to explore the ocean floor and overcome human limitations. Due to the programming of the robots, they can perform a more laborious task with extra hard work and with greater responsibility. Moreover, they do not wear out easily.

1.3 Digital Assistance Helps in Day to Day Chores

Siri listens to us and performs the task in one tap. GPS helps you to travel the world. How can I forget the basic necessity? Food, clothing, shelter, and smartphones. They are the ones that predict what we are going to type. In short, they know us better than anyone. The best is the autocorrect feature. It understands what you are trying to say and present you the sentence in the best way possible. Have you observed that while you post a picture on social media, you tag people, but the machine automatically detects the person's face and tags that individual? The same thing happens when you work on Google Photos. Automatically, a folder is created of the people with the help of their faces. Artificial intelligence is widely employed by financial institutions and banking institutions because it helps to organize and manage data. It is also used to detect fraud in a smart card-based system.

1.4 Rational Decision Maker

Logic above all! Highly advanced organizations have digital assistants which help them to interact with the users and save human resources.

Right program decisions can be made if they are worked upon rationally. But, with humans, emotions come along. For artificial thinkers, there is no distraction at all. They don't have any emotions, and that makes robots think logically. Thus they are always productive.

1.5 Repetitive Jobs

The same old talk goes a task that doesn't add value is of no use. Also, repetitive jobs are monotonous in nature and can be carried out with the help of machine intelligence. Machines think faster than humans and can perform various functions at the same time. It can be employed to carry out dangerous tasks, and its parameters are adjusted. This is not possible with humans as their speed and time can't be calculated on the basis of parameters.

1.6 Medical Applications

This is the best thing that artificial intelligence has done to humans. Doctors assess patients and their health risks with the help of artificial machine intelligence. The applications help to educate the machine about the side effects of various medicines. Nowadays, medical professionals are trained with artificial surgery simulators. It uses application which helps to detect and monitor neurological disorders and stimulate the brain functions. This also helps in the radiosurgery. Radiosurgery is used in operating tumors and it helps the operation without damaging the surrounding tissues.

1.7 Tireless, Selfless and with No Breaks

A machine doesn't require breaks like the way humans do. It is programmed for long hours and can continuously perform without getting bored or distracted. The machine does not get tired, even if it has to work for consecutive hours. This is a major benefit over humans, who need a rest from time to time to be efficient. However, in the case of machines, their efficiency is not affected by any external factor, and it does not get in the way of continuous work.

1.8 Right Decision Making

Because the machines have no emotions at all, they are able to make the right decisions in a short span of time. The best example of this is its usage in healthcare. The integration of AI tools in the healthcare sector has improved the efficiency of treatments by minimizing the risk of false diagnosis.

1.9 Implementing Artificial Intelligence in Risky Situations

Human safety is taken care of by machines. AI can be used in machines that are fitted with predefined algorithms. Scientists use complex machines to study the ocean floor where human survival becomes difficult. This is the level of AI. It reaches the place where humans can't reach. It helps to solve issues in a jiffy.

2. Disadvantages of Artificial Intelligence

2.1 High Cost

It's true that AI comes with a high cost. It requires huge costs as it is a complex machine. Apart from the installation cost, its repair and maintenance also require huge costs. The software programs need frequent upgradation to cater to the needs of the changing environment.

Also, if there is a breakdown, the cost of procurement is very high. With that, recovery requires huge time too.

2.2 No Human Replication

No matter how smart a machine becomes, it can never replicate a human. Machines are rational but very inhuman as they don't possess emotions and moral values. They don't know what is ethical and what's legal and because of this, they don't have their own judgment making skills. They do what they are told to do and therefore the judgment of right or wrong is nil for them. If they encounter a situation that is unfamiliar to them, they will perform incorrectly or else break down in such situations.

2.3 Creativity Is Not the Key for Artificial Intelligence

Machines can't be creative. They can only do what they are being taught or commanded. Though they help in designing and creating, they can't match the power of a human brain.

Humans are sensitive and intellectual, and they are very creative, too. They can generate ideas. They can think out of the box. Their thoughts are guided by the feelings which completely lacks in machines. No matter how much a machine outgrows, it can't inherent intuitive abilities of the human brain and can't replicate it.

2.4 Unemployment

This one is the riskiest and can have severe effects. With capital intensive technologies, human-intensive requirements have decreased in some industries. If human beings don't add to their skills in the future, we can see that they will be replaced with machines in no time.

New Words

ample	['æmpl]	adj.足够的，充足的
fulfill	[fʊl'fil]	vt.履行，执行；达到（目的）
robotic	[rəʊ'bɔtik]	adj.机器人的
limitation	[ˌlimi'teiʃn]	n.限制，局限
laborious	[lə'bɔ:riəs]	adj.费力的；勤劳的；辛苦的
autocorrect	['ɔ:təʊkə'rekt]	v.自动校正，自动更正
observe	[əb'zə:v]	v.观察；研究
folder	['fəʊldə]	n.文件夹
rational	['ræʃnəl]	adj.理性的；理智的
monotonous	[mə'nɔtənəs]	adj.单调的；枯燥无味的
dangerous	['deindʒərəs]	adj.危险的
parameter	[pə'ræmitə]	n.参数
adjust	[ə'dʒʌst]	v.适应，调整
simulator	['simjʊleitə]	n.模拟装置，模拟器
neurological	[ˌnjʊərə'lɔdʒikl]	adj.神经学的
disorder	[dis'ɔ:də]	n.混乱；（身心机能的）失调
radiosurgery	['reidiəʊ'sə:dʒəri]	n.放射外科
tireless	['taiəlis]	adj.不疲倦的，孜孜不倦的
continuously	[kən'tinjʊəsli]	adv.连续不断地
distract	[di'strækt]	vt.使分心；使混乱
consecutive	[kən'sekjʊtiv]	adj.连续的，连贯的
span	[spæn]	n.跨度 vt.跨越时间或空间
treatment	['tri:tmənt]	n.治疗，疗法；处理
predefined	[pri:di'faind]	v.预定义
survival	[sə'vaivl]	n.幸存，生存；幸存者
cost	[kɔst]	n.成本，代价
frequent	['fri:kwənt]	adj.频繁的，时常发生的，常见的
upgradation	[ˌʌpgrei'deiʃn]	n.升级；改善；提高
breakdown	['breikdaʊn]	n.崩溃；损坏，故障
procurement	[prə'kjʊəmənt]	n.采购；获得，取得
recovery	[ri'kʌvəri]	n.恢复，复原

inhuman	[in'hju:mən]	adj.非人的，不近人情的
encounter	[in'kauntə]	vt.不期而遇；遭遇；对抗
incorrectly	[ˌinkə'rektli]	adv.不正确地，错误地
command	[kə'mɑ:nd]	n.命令
sensitive	['sensitiv]	adj.敏感的；感觉的
intellectual	[ˌintə'lektʃuəl]	adj.智力的；有才智的
outgrow	[ˌaut'grəu]	vt.过度成长
inherent	[in'hiərənt]	adj.固有的，内在的
intuitive	[in'tju:itiv]	adj.直觉的；直观的
replace	[ri'pleis]	vt.替换，代替

Phrases

digital assistant	数字助理，数字助手
wear out	用坏；耗尽
financial institution	金融机构
banking institution	银行机构
smart card-based system	基于智能卡的系统
human resource	人力资源
carry out	实行，被执行
side effect	（药物等起到的）副作用
external factor	外部因素，外界因素
moral value	道德价值，道德观
think out of the box	打破陈规

Text B 参考译文
人工智能的利与弊

1. 人工智能的优势

1.1 是人都会"犯错"，那么为什么不使用 Artificial Intelligence

机器根据先前的数据记录做出决策。使用算法可以减少出错的机会，这是一项大成就，因为解决复杂的问题需要艰难且无误的计算。商业组织使用数字助理与用户进行交互，这可以帮助他们节省大量时间，满足用户业务的需求，因此他们不必等待。数字助理经过编程，可以为用户提供最佳的帮助。

1.2 Artificial Intelligence 不会疲倦也不易损坏

人工智能和机器人科学被用于采矿和其他燃料勘探过程，这些复杂的机器有助于探索海底并克服人类的局限性。由于对机器人进行了编程，它们可以执行更费力的任务，完成更艰苦的工作和担负更大的责任。而且，它们不容易损坏。

1.3 数字助理可帮助你处理日常琐事

Siri 听我们的指令并能马上执行任务，GPS 可帮助你环游世界。我怎么会忘记基本需求？食物、衣服、住所和智能手机。数字助理可以预测我们将要键入的内容，总之，它们比任何人都更了解我们。自动更正功能尤为出色，它可以理解你要说的内容，并以最佳方式向你展示句子。你是否观察到在社交媒体上发布图片时会标记人物，但是机器会自动检测人物的脸并标记人物？使用 Google 相册时也是如此，它会自动根据人们的面容创建一个文件夹。人工智能在金融机构和银行机构广泛使用，因为它有助于组织和管理数据。同样，人工智能也可用来检测基于智能卡系统中的欺诈。

1.4 理性决策者

逻辑高于一切！高度先进的组织拥有数字助理，可以帮助组织与用户互动并节省人力资源。通过理性思考就可以做出正确的程序决策，但是，人是有情感的。对于人工思考者来说，它们根本不会分心，它们没有任何情感。这让机器人具有逻辑思维，因此，它们总是富有成效的。

1.5 重复性工作

老话说，没有增加价值的劳作是无效的。重复性工作本质上是单调的，因而可以借助机器智能来执行。机器的思维速度比人类快，并且可以同时执行各种功能。它可以用于执行危险任务，并且可以调整任务参数。对于人类而言，这是不可能的，因为无法根据参数来计算他们的速度和时间。

1.6 医疗应用

这是人工智能对人类所做的最好的事情，医生借助人工智能来评估患者及其健康风险。这些应用程序有助于教育机器有关各种药物的副作用。如今，医学专业人员接受了人工手术模拟器的培训，这种培训使用有助于检测和监测神经系统疾病并刺激大脑功能的应用程序，这对放射外科手术很有帮助。放射外科手术用于治疗肿瘤，人工智能辅助手术而不损害周围组织。

1.7 孜孜不倦，无私无休

机器不像人类一样需要休息，它可以长时间、连续运行而不会感到无聊或分心，即使必须连续工作几个小时，它也不会疲劳。与人类相比，这是一个重要优点，因为人类需要时不时地休息以提高效率。但是，对于机器而言，它们的效率不受任何外部因素的影响，即使连续工作也无妨。

1.8 正确的决策

由于机器完全没有情绪，因此它们能够在短时间内做出正确的决定，最好的例子是其在医疗保健中的使用。通过将 AI 工具集成到医疗保健行业中，可以最大限度地减少错误诊断的风险，从而提高治疗效率。

1.9 在有风险的情况下使用 Artificial Intelligence

人身安全由机器负责，在装有预定义算法的机器上都可以使用 AI，科学家使用复杂的机器研究人类难以生存的海底。这就是 AI 的能力，它可以到达人类无法到达的地方，有助于即刻解决问题。

2. 人工智能的劣势

2.1 高成本

的确，AI 的成本很高。它是一种复杂的机器，因此需要巨额成本。除了安装成本外，其维修和保养也需要巨额成本。软件程序需要经常升级，以适应不断变化的环境的需求。

另外，如果发生故障，相关采购成本也很高，而且，恢复也需要大量时间。

2.2 不能复制人类

不管机器变得多么聪明，它都无法复制人类。机器是理性的，不具备任何人性，因为它们没有情感和道德价值。它们不知道什么是道德和法律，因此，它们没有自己的判断能力。它们按照指示去做，因此它们不能判断是非，如果遇到不熟悉的情况，则不能正确运行，甚至会在这种情况下崩溃。

2.3 创造力不是 Artificial Intelligence 的强项

机器没有创造力，它们只能做自己被教导或命令的事情，尽管它们有助于设计和创建，但它们无法匹敌人脑的力量。

人类是敏感和有智力的，而且也很有创造力，可以产生想法，可以打破陈规。人的思想受感情的指导而机器完全缺乏感觉。不管机器发展得多厉害，它都不会具有人脑固有的直觉能力，也无法复制这种能力。

2.4 失业

这是最危险的，可能会造成严重的后果。随着资本密集型技术的发展，某些行业中对人力密集型的需求有所降低。如果将来人类不再增加自己的技能，那么我们很快就会看到他们将被机器取代。

Exercises

[Ex. 1] Answer the following questions according to Text A.

1. What is the main aim of artificial intelligence? What is the primary way of classifying AI based on?
2. What is the most basic and oldest type of artificial intelligence? Why does it lack the capability to use previously gained information/experience to obtain better results?
3. What do most of the common applications existing around us fall under? How can these AI applications be trained?
4. What is said about theory of mind?
5. What can self-aware AI do?
6. What does ANI include? Why do they fail at performing unprecedented task?
7. What capability does AGI have?
8. Why will ASI be able to perform all the tasks better than humans?
9. What is technological singularity?
10. What can we say for the proponents of AI? And what about for the AI skeptics?

[Ex. 2] Fill in the following blanks according to Text B.

1. Machines make decision based on _____. With algorithms, the chances of errors _____.
2. Due to the programming of the robots, they can perform _____ with extra hard work and with _____. Moreover, they _____.
3. Artificial intelligence is widely employed by _____ and _____ because it helps to organize and manage data. It is also used to _____.
4. Highly advanced organizations have digital assistants which help them to _____ and _____. For artificial thinkers, there is _____. They don't have any

emotions, and that makes robots _____. Thus they are always _____.
5. Nowadays, medical professionals are trained with _____. It uses application which helps to detect and monitor _____ and stimulate _____.
6. Because the machines have no emotions at all, they are able to _____ in a short span of time. The best example of this is its usage in _____.
7. Scientists use complex machines to study the ocean floor where _____. This is the level of AI. It reaches the place where _____. It helps to solve issues _____.
8. It's true that AI comes with _____. It requires huge costs as it is _____. Apart from the installation cost, _____ also require huge costs.
9. Machines are rational but _____ as they don't possess emotions and _____. They don't know _____ and _____ and because of this, they don't have their own judgment making skills.
10. Machines can't be creative. They can only do _____ or _____. Though they help in _____, they can't match the power of a human brain.

[Ex. 3] Translate the following terms or phrases from English into Chinese and vice versa.

1. prediction modelling   1. _____
2. reactive machine   2. _____
3. reference model   3. _____
4. theory of mind   4. _____
5. common sense   5. _____
6. n.类型，种类，类别   6. _____
7. n.经验，体验；经历   7. _____
8. n.功能，功能性   8. _____
9. adj.潜在的，有可能的   9. _____
10. n.原型，雏形，蓝本   10. _____

[Ex. 4] Translate the following passages into Chinese.

Pros of Artificial Intelligence

1. Integration with Other Technologies

The most positive impact utilizing AI can be achieved by integration with other technologies. For instance, both IoT and AI need to work in conjunction to enable self-driving cars. Where IoT is responsible for activating and regulating sensors in the car that collect real-time data, vehicle relies on

Unit 2 Types, Pros and Cons of Artificial Intelligence

AI for decision making. Similarly, integration with Blockchain would enable better security and scalability.

2. Less Reliance on Traditional Industries

As more and more companies realise the benefits of AI and incorporate it into their operations, the traditional industries should become less relevant. For instance, with self-driving cars, the traditional players would either have to adapt or become irrelevant.

3. Enabling Real-Time Customer Interactions

With more AI driven real-time data and activities, customer interactions across channels will become truly real-time. AI marketing will help with customer retention, winning customers back and engaging first timers: Enabled by targeting social media and platform campaigning.

4. Automation of Processes

AI has been drastically adopted to automate key processes across sectors such as retail, banking, and manufacturing. And it is predicted that more will follow.

5. Utilization of Artificial Intelligence Assistants

AI assistants have been successfully installed to streamline and automate customer service and sales tasks. Some of the most common ones are: Siri, Cortana, Alexa and Google Assistant. And it is expected that more companies will opt for AI assistants to handle basic tasks.

6. Elimination of Biased Data

Organisations have relied on Machine Learning models for critical decisions like hiring or loan approvals. Since biased data is an inherent risk with Machine Learning, AI based applications are going to be useful in making more informed decisions.

7. Facial Recognition Adoption

One way to ensure biometric authentication is through facial recognition. Investments and researches have been directed at improving AI applications' accuracy and readability. In 2022, it is expected that AI facial recognition technology will be used widely.

[Ex. 5] Fill in the blanks with the words given below.

    rules   determining   simulation   database   conversation computer   cognitive   complicated   predefined   narrow

Weak Artificial Intelligence

Weak artificial intelligence (weak AI) is an approach to artificial intelligence research and development with the consideration that AI is and will always be a ___1___ of human cognitive function, and that computers can only appear to think but are not actually conscious in any sense of the word. Weak AI simply acts upon and is bound by the ___2___ imposed on it, and it could not go beyond those rules. A good example of weak AI are characters in a ___3___ game that act believably within the context of their game character, but are unable to do anything beyond that.

Weak artificial intelligence is also known as narrow artificial intelligence.

Weak artificial intelligence is a form of AI specifically designed to be focused on a ___4___ task and to seem very intelligent at it. It contrasts with strong AI, in which an AI is capable of all and any ___5___ functions that a human may have, and is in essence no different than a real human mind. Weak AI is never taken as a general intelligence but rather a construct designed to be intelligent in the narrow task that it is assigned to.

A very good example of a weak AI is Apple's Siri, which has the Internet behind it serving as a powerful ___6___. Siri seems very intelligent, as it is able to hold a ___7___ with actual people, even giving snide remarks and a few jokes, but actually operates in a very narrow, ___8___ manner. However, the "narrowness" of its function can be evidenced by its inaccurate results when it is engaged in conversations that it is not programmed to respond to.

Robots used in the manufacturing process can also seem very intelligent because of the accuracy and the fact that they are doing very ___9___ actions that could seem incomprehensible to a normal human mind. But that is the extent of their intelligence; they know what to do in the situations that they are programmed for, and outside of that they have no way of ___10___ what to do. Even AI equipped for machine learning can only learn and apply what it learns to the scope it is programmed for.

Unit 3
Knowledge Representation and Reasoning

Text A
Knowledge Representation in Artificial Intelligence

扫码听课文

Human are best at understanding, reasoning, and interpreting knowledge. Human knows things, which is knowledge and as per their knowledge they perform various actions in the real world. But how machines do all these things comes under knowledge representation and reasoning. Hence we can describe knowledge representation as following:

- Knowledge representation and reasoning is the part of artificial intelligence which is concerned with AI agent's thinking and how thinking contributes to intelligent behavior of agents.
- It is responsible for representing information about the real world so that a computer can understand and utilize this knowledge to solve the complex real world problems such as diagnosing medical condition or communicating with humans in natural language.
- It is also a way which describes how we can represent knowledge in artificial intelligence. Knowledge representation is not just storing data into some database, but it also enables an intelligent machine to learn from that knowledge and experiences so that it can behave intelligently like a human.

1. What to Represent

The following are the kind of knowledge which needs to be represented in AI systems:
- Object: All the facts about objects in our world domain. E.g., Guitars contain strings,

trumpets are brass instruments.
- Events: Events are the actions which occur in our world.
- Performance: It describe behavior which involves knowledge about how to do things.
- Meta-knowledge: It is knowledge about what we know.
- Facts: Facts are the truths about the real world, and what we represent.
- Knowledge-Base: The central component of the knowledge-based agents is the knowledge base. It is represented as KB. The knowledge base is a group of the sentences (Here, sentences are used as a technical term and not identical with the English language).
- Knowledge: Knowledge is awareness or familiarity gained by experiences of facts, data, and situations.

2. Types of Knowledge

The following are the various types of knowledge (see Figure 3-1):

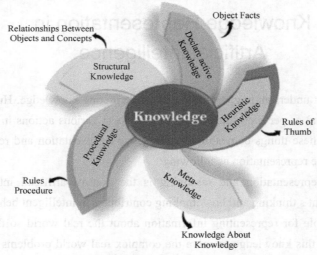

Figure 3-1　Types of knowledge

2.1　Declarative Knowledge

Declarative knowledge is to know about something. It includes concepts, facts, and objects. It is also called descriptive knowledge and expressed in declarative sentences. It is simpler than procedural language.

2.2　Procedural Knowledge

It is also known as imperative knowledge. Procedural knowledge is a type of knowledge which is responsible for knowing how to do something. It can be directly applied to any task. It includes rules, strategies, procedures, agendas, etc. Procedural knowledge depends on the task to which it can be applied.

2.3 Meta-Knowledge

Knowledge about the other types of knowledge is called meta-knowledge.

2.4 Heuristic Knowledge

Heuristic knowledge is representing knowledge of some experts in a field or subject. It is rules of thumb based on previous experiences, awareness of approaches, and which are good to work but not guaranteed.

2.5 Structural Knowledge

Structural knowledge is basic knowledge to problem-solving. It describes relationships between various concepts such as kind of, part of, and grouping of something. It describes the relationships that exist between concepts or objects.

3. The Relation Between Knowledge and Intelligence

Knowledge of real world plays a vital role in intelligence and the same for creating artificial intelligence. Knowledge plays an important role in demonstrating intelligent behavior in AI agents. An agent is only able to accurately act on some input when it has some knowledge or experience about that input.

Let's suppose if you meet some person who is speaking in a language which you don't know, then how you will be able to act on that. The same thing applies to the intelligent behavior of the agents.

As we can see in the diagram below (see Figure 3-2), there is one decision maker which acts by sensing the environment and using knowledge. But if the knowledge part does not present then, it cannot display intelligent behavior.

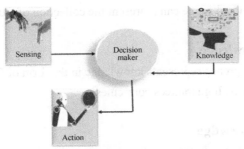

Figure 3-2 The relation between knowledge and intelligence

4. Artificial Intelligence Knowledge Cycle

An artificial intelligence system has the following components for displaying intelligent behavior: perception, learning, knowledge representation, and reasoning, planning, and execution.

AI system has perception component by which it retrieves information from its environment. It can be visual, audio or another form of sensory input. The learning component is responsible for learning from data captured by perception comportment. In the complete cycle, the main components are knowledge representation and reasoning. These two components are independent with each other but also coupled together. The planning and execution depend on analysis of knowledge representation and reasoning.

5. Approaches to Knowledge Representation

There are mainly four approaches to knowledge representation.

5.1 Simple Relational Knowledge

It is the simplest way of storing facts which uses the relational method, and each fact about a set of the objects is set out systematically in columns. This approach of knowledge representation is famous in database systems where the relationship between different entities is represented. This approach has little opportunity for inference.

5.2 Inheritable Knowledge

In the inheritable knowledge approach, all data must be stored into a hierarchy of classes. All classes should be arranged in a generalized form or a hierarchical manner. In this approach, we apply inheritance property. Elements inherit values from other members of a class. Objects and values are represented in Boxed nodes. We use Arrows which point from objects to their values. This approach contains inheritable knowledge which shows a relation between instance and class, and it is called instance relation. Every individual frame can represent the collection of attributes and its value.

5.3 Inferential Knowledge

Inferential knowledge approach represents knowledge in the form of formal logic. This approach can be used to derive more facts. It guarantees correctness.

5.4 Procedural Knowledge

Procedural knowledge approach uses small programs and codes to describe how to do specific things, and how to proceed. In this approach, one important rule is used, which is If-Then rule. In this knowledge, we can use various coding languages such as LISP language and Prolog language. We can easily represent heuristic or domain-specific knowledge using this approach. But it is not necessary that we can represent all cases in this approach.

6. Requirements for Knowledge Representation System

A good knowledge representation system must possess the following properties.

Representational accuracy: KR system should have the ability to represent all kind of required knowledge.

Inferential adequacy: KR system should have ability to manipulate the representational structures to produce new knowledge corresponding to existing structure.

Inferential efficiency: The ability to direct the inferential knowledge mechanism into the most productive directions by storing appropriate guides.

Acquisitional efficiency: The ability to acquire the new knowledge easily using automatic methods.

New Words

knowledge	['nɔlidʒ]	n. 知识；了解，理解
representation	[ˌreprizen'teiʃn]	n. 表示，表述
reasoning	['ri:zəniŋ]	n. 推理，论证
		adj. 推理的
interpret	[in'tə:prit]	vt. 解释；理解
action	['ækʃn]	n. 行动，活动
contribute	[kən'tribju:t]	v. 贡献出；出力
database	['deitəbeis]	n. 数据库；资料库；信息库
behave	[bi'heiv]	vi. 表现；自然反应
object	['ɔbdʒikt]	n. 对象；物体
domain	[də'mein]	n. 范围，领域
event	[i'vent]	n. 事件
behavior	[bi'heivjə]	n. 行为
meta-knowledge	['metə'nɔlidʒ]	n. 元知识
knowledge-based	['nɔlidʒ beist]	adj. 基于知识的

英文	音标	释义
identical	[aɪˈdentɪkl]	adj.同一的；相同的
awareness	[əˈweənɪs]	n.察觉，觉悟，意识
familiarity	[fəˌmɪliˈærəti]	n.熟悉；通晓；认识
situation	[ˌsɪtʃʊˈeɪʃn]	n.情境，情况
declarative	[dɪˈklærətɪv]	adj.陈述的，宣言的，公布的
concept	[ˈkɔnsept]	n.概念
procedural	[prəˈsiːdʒərəl]	adj.程序的
imperative	[ɪmˈperətɪv]	adj.必要的；命令的
strategy	[ˈstrætɪdʒi]	n.策略，战略
agendas	[əˈdʒendəz]	n.议事日程(agenda 的名词复数)
heuristic	[hjʊəˈrɪstɪk]	adj.启发式的；探试的，探索的
structural	[ˈstrʌktʃərəl]	adj.结构化的
relationship	[rɪˈleɪʃnʃɪp]	n.关系，联系
input	[ˈɪnpʊt]	n.&v.输入
diagram	[ˈdaɪəɡræm]	n.图表；示意图
		vt.用图表示；图解
capture	[ˈkæptʃə]	vt.捕获，捕捉
comportment	[kəmˈpɔːtmənt]	n.行动，行为
independent	[ˌɪndɪˈpendənt]	adj.独立的，自主的；不相关连的
systematically	[ˌsɪstɪˈmætɪkli]	adv.有系统地；有组织地
entity	[ˈentəti]	n.实体
inference	[ˈɪnfərəns]	n.推理；推断；推论
inheritable	[ɪnˈherɪtəbl]	adj.可继承的，会遗传的
hierarchy	[ˈhaɪərɑːki]	n.分层，层次
generalized	[ˈdʒenrəlaɪzd]	adj.广泛的；普遍的
manner	[ˈmænə]	n.方式，方法
property	[ˈprɔpəti]	n.特性；属性
inherit	[ɪnˈherɪt]	v.继承
node	[nəʊd]	n.节点
frame	[freɪm]	n.框架
guarantee	[ˌɡærənˈtiː]	vt.保证，担保
correctness	[kəˈrektnɪs]	n.正确性
possess	[pəˈzes]	vt.拥有；掌握，懂得
representational	[ˌreprɪzenˈteɪʃnl]	adj.代表性的

manipulate	[mə'nipjʊleit]	vt.	操作，处理
correspond	[ˌkɔri'spɔnd]	vi.	符合，一致；相应
mechanism	['mekənizəm]	n.	机制
acquisitional	[ˌækwi'ziʃənəl]	adj.	习得的

Phrases

be concerned with	涉及；与……有关
real world	真实世界
be responsible for ...	为……负责，是造成……的原因
communicate with ...	与……交流
natural language	自然语言
declarative knowledge	陈述性知识
procedural knowledge	程序知识
imperative knowledge	命令式知识
heuristic knowledge	启发式知识
structural knowledge	结构化知识
act on ...	对……起作用，按照……而行动
simple relational knowledge	简单关系知识
set out	着手；安排
formal logic	形式逻辑

Abbreviations

KR (Knowledge Representation)	知识表述
KRR (Knowledge Representation and Reasoning)	知识表述及推理
KB (Knowledge Base)	知识库

Text A 参考译文
人工智能中的知识表示

人类最擅长知识的理解、推理和解释。人类了解事物，这就是知识，根据所掌握的知识，他们在现实世界中执行各种动作。但是机器要做这些事情就取决于知识的表达和推理，因此，我们可以将知识表述描述如下：
- 知识表述和推理是人工智能的一部分，它与人工智能代理的思维以及思维如何促进代理的智能行为有关。
- 它负责表述有关现实世界的信息，以便计算机可以理解并可以利用这些知识来解决复杂的现实世界中的问题，例如诊断医疗状况或以自然语言与人交流。
- 这也是描述我们如何表示人工智能知识的一种方式。知识表述不仅将数据存储到某个数据库中，而且还使智能机器能够从该知识和经验中学习，从而可以像人一样智能地运行。

1. 表述什么

以下是需要在人工智能系统中表述的知识类型：
- 对象：我们世界范围内有关对象的所有事实。例如，吉他包含弦，小号是铜管乐器。
- 事件：即我们世界中发生的行为。
- 表现：描述行为，涉及有关如何做事的知识。
- 元知识：这是关于我们所知道的知识。
- 事实：事实是关于现实世界和我们所表述的真相。
- 知识库：基于知识代理的核心部件是知识库，它表示为KB。知识库是一组语句（此处的语句用作技术术语，与英语语句不相同）。
- 知识：知识是通过事实、数据和情境的经验获得的意识或认识。

2. 知识类型

以下是各种类型的知识（见图3-1）：

（图略）

2.1 陈述性知识

陈述性知识是将要了解的一些东西，它包括概念、事实和对象。它也被称为描述性知识，并以陈述性语句来表示，比过程语言更简单。

2.2 程序知识

它也称为命令式知识，涉及了解如何做某事，可以直接应用于任何任务，包括规则、策略、程序、议程等。程序知识取决于它可以应用的任务。

2.3 元知识

关于其他类型知识的知识称为元知识。

2.4 启发性知识

启发性知识表示一个领域或学科中某些专家的知识，它是基于以前的经验、对方法的了解以及哪些方法会更好，但不能保证一定有效的经验法则。

2.5 结构化知识

结构化知识是解决问题的基础知识。它描述了各种概念之间的关系，例如某种事物的种类、部分和分组。它描述了概念或对象之间存在的关系。

3. 知识与智力的关系

现实世界的知识在智能中起着至关重要的作用，在创造人工智能方面也起着至关重要的作用。知识在显示人工智能代理中的智能行为方面起着重要作用。只有对某项输入有一定知识或经验时，代理才能够准确地对某项输入采取行动。

让我们假设，如果你遇到某个人，你不懂他的语言，那么你对他说的话会做出什么样的反应？同样的情况也适用于代理的智能行为。

正如我们在下图中所看到的（见图 3-2），一个决策者通过感知环境和使用知识来行动。但是，如果知识部分不存在，那么它将无法显示智能行为。

（图略）

4. 人工智能知识周期

人工智能系统具有以下用于显示智能行为的部分：感知、学习、知识表述和推理、计划和执行。

人工智能系统具有感知部分，可通过它从环境中检索信息，可以是视频、音频或其他形式的感觉输入。学习部分负责从感知行动中捕获的数据中学习。在整个周期中，主要组成部分是知识表述和推理，这两个部分彼此独立，但也耦合在一起。计划和执行取决于对知识表述和推理的分析。

5. 知识表述方法

知识表述主要有以下四种方法：

5.1 简单的关系知识

这是使用关系方法存储事实的最简单方法，有关一组对象的每个事实都系统地列出。这种知识表述方法在表示不同实体之间关系的数据库系统中很有名。这种方法几乎没有推理的机会。

5.2 可继承的知识

在可继承知识方法中，所有数据必须存储在类的层次结构中，所有类都应以广义形式或分层方式排列。在这种方法中，我们应用继承属性。元素从类的其他成员继承值。对象和值以加框节点表示，用箭头表示从对象到值。这种方法包含可继承的知识，该知识显示了实例与类之间的关系，称为实例关系。每个单独的框架都可以代表属性及其值的集合。

5.3 推理知识

推理知识方法以形式逻辑的形式表述知识，此方法可用于得出更多事实，它保证了知识的正确性。

5.4 程序知识

程序知识方法使用小的程序和代码来描述如何做特定的事情以及如何进行。在这种方法中，使用了一个重要规则，即 If-Then 规则。据此，我们可以使用各种编码语言，例如 LISP 语言和 Prolog 语言。使用这种方法，我们可以轻松地表述启发性或特定领域的知识，但是我们不能用这种方法表述所有情况。

6. 对知识表述系统的要求

一个好的知识表述系统必须具有以下特性。

表述准确性：知识表述系统应具有表述所有必需知识的能力。

推论充足：知识表述系统应具有处理表述结构的能力，以产生与现有结构相对应的新知识。

推理效率：能够通过存储适当的指南，将推理知识机制引导到最具生产力的方向。

习得效率：能够使用自动方法轻松掌握新知识。

Text B
Reasoning in Artificial Intelligence

扫码听课文

Reasoning is the mental process of deriving logical conclusion and making predictions from available knowledge, facts, and beliefs. Or we can say, "Reasoning is a way to infer facts from existing data." It is a general process of thinking rationally to find valid conclusions.

In artificial intelligence, reasoning is essential so that the machine can also think rationally as a human brain, and can perform like a human.

1. Deductive Reasoning

Deductive reasoning is deducing new information from logically related known information. It is the form of valid reasoning, which means the argument's conclusion must be true when the premises are true.

Deductive reasoning is a type of propositional logic in AI, and it requires various rules and facts. It is sometimes referred to as top-down reasoning, and contradictory to inductive reasoning.

In deductive reasoning, the truth of the premises guarantees the truth of the conclusion.

Deductive reasoning mostly starts from the general premises to the specific conclusion, which can be explained as the below example.

Example:

Premise-1: All the human eats veggies

Premise-2: Suresh is human.

Conclusion: Suresh eats veggies.

The general process of deductive reasoning is given below:

$$\text{Theory} \rightarrow \text{Hypothesis} \rightarrow \text{Patterns} \rightarrow \text{Confirmation}$$

2. Inductive Reasoning

Inductive reasoning is a form of reasoning to arrive at a conclusion using limited sets of facts by the process of generalization. It starts with a series of specific facts or data and reaches to a general statement or conclusion.

Inductive reasoning is a type of propositional logic, which is also known as cause-effect reasoning or bottom-up reasoning.

In inductive reasoning, we use historical data or various premises to generate a generic rule, for which premises support the conclusion.

In inductive reasoning, premises provide probable supports to the conclusion, so the truth of premises does not guarantee the truth of the conclusion.

Example:

Premise: All of the pigeons we have seen in the zoo are white.

Conclusion: Therefore, we can expect all the pigeons to be white.

The general process of inductive reasoning is given below:

$$\text{Observation} \rightarrow \text{Patterns} \rightarrow \text{Hypothesis} \rightarrow \text{Theory}$$

3. Abductive Reasoning

Abductive reasoning is a form of logical reasoning which starts with a single or multiple observations, then seeks to find the most likely explanation or conclusion for the observation.

Abductive reasoning is an extension of deductive reasoning, but in abductive reasoning, the premises do not guarantee the conclusion.

Example:
Implication: Cricket ground is wet if it is raining
Axiom: Cricket ground is wet.
Conclusion: It is raining.

4. Common Sense Reasoning

Common sense reasoning is an informal form of reasoning, which can be gained through experiences.

Common sense reasoning simulates the human ability to make presumptions about events which occurs on every day.

It relies on good judgment rather than exact logic and operates on heuristic knowledge and heuristic rules.

Example:
One person can be at one place at a time.
If I put my hand in a fire, then it will burn.

The above two statements are the examples of common sense reasoning which a human mind can easily understand and assume.

5. Monotonic Reasoning

In monotonic reasoning, once the conclusion is made, it will remain the same even if we add some other information to the existing information in our knowledge base. In monotonic reasoning, adding knowledge does not decrease the set of propositions that can be derived.

To solve monotonic problems, we can derive the valid conclusion from the available facts only, and it will not be affected by new facts.

Monotonic reasoning is not useful for the real-time systems, as in real time, facts get changed, so we cannot use monotonic reasoning.

Monotonic reasoning is used in conventional reasoning systems, and a logic-based system is monotonic.

Any theorem proving is an example of monotonic reasoning.

Example:

Earth revolves around the Sun.

It is a true fact, and it cannot be changed even if we add another sentence in knowledge base like, "The moon revolves around the earth" or "Earth is not round," etc.

Advantages of monotonic reasoning:

In monotonic reasoning, each old proof will always remain valid.

If we deduce some facts from available facts, then it will remain valid for always.

Disadvantages of monotonic reasoning:

We cannot represent the real-world scenarios using monotonic reasoning.

Hypothesis knowledge cannot be expressed with monotonic reasoning, which means facts should be true.

Since we can only derive conclusions from the old proofs, new knowledge from the real world cannot be added.

6. Non-monotonic Reasoning

In non-monotonic reasoning, some conclusions may be invalidated if we add some more information to our knowledge base.

Logic will be said as non-monotonic if some conclusions can be invalidated by adding more knowledge into our knowledge base.

Non-monotonic reasoning deals with incomplete and uncertain models.

"Human perceptions for various things in daily life," is a general example of non-monotonic reasoning.

Example: Let's suppose the knowledge base contains the following knowledge:

Birds can fly

Penguins cannot fly

Pitty is a bird

So from the above sentences, we can conclude that Pitty can fly.

However, if we add one another sentence into knowledge base "Pitty is a penguin", which concludes "Pitty cannot fly", so it invalidates the above conclusion.

Advantages of non-monotonic reasoning:

For real-world systems such as robot navigation, we can use non-monotonic reasoning.

In non-monotonic reasoning, we can choose probabilistic facts or can make assumptions.

Disadvantages of non-monotonic reasoning:

In non-monotonic reasoning, the old facts may be invalidated by adding new sentences. It cannot be used for theorem proving.

New Words

mental	['mentl]	adj.思考的；智慧的，智力的
belief	[bi'li:f]	n.信念，信条
infer	[in'fə:]	vt.推断；猜想
rationally	['ræʃnəli]	adv.讲道理地，理性地
valid	['vælid]	adj.有效的；正当的
essential	[i'senʃl]	adj.基本的；必要的；本质的
deductive	[di'dʌktiv]	adj.演绎的
argument	['ɑ:gjʊmənt]	n.论点；论据
premise	['premis]	n.前提
propositional	[prɔpə'ziʃənəl]	adj.命题的
contradictory	[ˌkɔntrə'diktəri]	adj.对立的，矛盾的
hypothesis	[hai'pɔθisis]	n.假设，假说；前提
confirmation	[ˌkɔnfə'meiʃn]	n.确认，认可；证实；证明
inductive	[in'dʌktiv]	adj.归纳的
generalization	[ˌdʒenrəlai'zeiʃn]	n.归纳；一般化；普通化
probable	['prɔbəbl]	adj.可能的，大概的
abductive	[æb'dʌktiv]	adj.溯因的
explanation	[ˌeksplə'neiʃn]	n.解释，说明
extension	[ik'stenʃn]	n.扩展，伸展，扩大
informal	[in'fɔ:ml]	adj.非正式的
simulate	['simjʊleit]	vt.模仿，模拟
exact	[ig'zækt]	adj.准确的；严密的；精密的，精确的
assume	[ə'sju:m]	v.假定，认为
remain	[ri'mein]	vi.留下；保持
preposition	[ˌprepə'ziʃn]	n.介词；前置词
affect	[ə'fekt]	vt.影响
conventional	[kən'venʃnəl]	adj.传统的；平常的
theorem	['θiərəm]	n.公理，定律，法则

proof	[pru:f]	n.证明
represent	[ˌrepri'zent]	vt.表现，代表，代理
scenario	[si'nɑ:riəʊ]	n.情景，情节；事态
invalidate	[in'vælideit]	vt.使无效；使作废；证明……错误
uncertain	[ʌn'sə:tn]	adj.不确定的；不明确的
assumption	[ə'sʌmpʃn]	n.假定，假设
sentence	['sentəns]	n.句子；命题；判定

Phrases

propositional logic	命题逻辑
top-down reasoning	自上而下推理
inductive reasoning	归纳推理
cause-effect reasoning	因果推理
bottom-up reasoning	自下而上推理
generic rule	通用规则，普通规则
abductive reasoning	溯因推理
common sense reasoning	常识推理
monotonic reasoning	单调推理
real-time system	实时系统
logic-based system	基于逻辑的系统
non-monotonic reasoning	非单调推理
deal with	处理；涉及
theorem proving	定理证明

Text B 参考译文
人工智能中的推理

推理是从逻辑上得出结论并根据已有的知识、事实和信条做出预测的思考过程。或者我们可以说："推理是从现有数据推断事实的一种方式。"它是寻求有效结论的理性思考的一般过程。

在人工智能中，推理至关重要，因此机器还可以像人的大脑一样理性地思考，并可以像人一样地工作。

1. 演绎推理

演绎推理是从逻辑上相关的已知信息中推论出新信息。这是有效推理的形式,这意味着当前提为真时,论点的结论必定为真。

演绎推理是人工智能中的一种命题逻辑,它需要各种规则和事实。它有时被称为自上而下的推理,与归纳推理相对应。

在演绎推理中,前提的真实性保证了结论的真实性。

演绎推理主要从一般前提到具体结论,可以通过下面的例子加以解释。

例:

前提 1:所有人都吃蔬菜

前提 2:Suresh 属于人类。

结论:Suresh 吃蔬菜。

演绎推理的一般过程如下:

理论→假设→模式→确认

2. 归纳推理

归纳推理是通过概括过程使用有限的事实集得出结论的一种推理形式,它从一系列特定的事实或数据开始,直至得出一般性陈述或结论。

归纳推理是一种命题逻辑,也称为因果推理或自下而上的推理。

在归纳推理中,我们使用历史数据或各种前提来生成通用规则,前提支持结论。

在归纳推理中,前提为结论提供可能的支持,因此前提的真实性不能保证结论的真实性。

例:

前提:我们在动物园看到的鸽子都是白色的。

结论:因此,我们可以认为所有鸽子都是白色的。

归纳推理的一般过程如下:

观察→模式→假设→理论

3. 溯因推理

溯因推理是一种逻辑推理的形式,其始于单一或多项观察,然后试图找到最可能的解释或结论。

溯因推理是演绎推理的扩展,但是在溯因推理中,前提并不能保证结论。

例:

含义:如果正在下雨,板球场是湿的。

公理：板球场是湿的。
结论是下雨了。

4. 常识推理

常识推理是一种非正式的推理形式，可以通过经验获得。
常识推理模拟了人类对每天发生的事件做出假设的能力。
它依赖于良好的判断力而不是确切的逻辑，并且基于启发性知识和启发性规则进行操作。
例：
一个人在某时只能在一个地方。
如果我将手放在火中，那么它将会烧伤。
以上两个陈述是人脑可以轻松理解和假定的常识推理示例。

5. 单调推理

在单调推理中，一旦得出结论，即使将一些其他信息添加到知识库中的现有信息中，结论也将保持不变。在单调推理中，添加知识不会减少可派生的命题集。
要解决单调推理问题，我们只能从现有事实中得出有效结论，结论不会受到新事实的影响。
单调推理对实时系统没有用，因为实时情况下事实会发生变化，因此我们不能使用单调推理。
在传统的推理系统中使用单调推理，而基于逻辑的系统是单调的。
任何定理证明都是单调推理的一个例子。
例：
地球围绕太阳旋转。
这是事实，即使在知识库中添加"月亮绕地球旋转"或"地球不是圆的"之类的知识，也无法更改该事实。
单调推理的优点：
在单调推理中，每个已有的证明将始终保持有效。
如果我们从现有事实中推断出一些事实，那么它将永远有效。
单调推理的缺点：
我们不能使用单调推理来展示现实世界场景。
假设的知识不能用单调推理来表示，这意味着事实应该是真实的。
由于我们只能从已有的证据中得出结论，因此无法添加来自现实世界的新知识。

6. 非单调推理

在非单调推理中，如果我们在知识库中添加更多信息，则某些结论可能无效。
如果通过在我们的知识库中添加更多知识可以使某些结论无效，那么其中的逻辑将被称为非单调的。

非单调推理处理不完整和不确定的模型。

"人们对日常生活中各种事物的看法"是非单调推理的一般示例。

示例：假设该知识库包含以下知识：

鸟儿会飞

企鹅不能飞

皮蒂是一只鸟

那么，从以上句子中我们可以得出结论，皮蒂可以飞。

但是，如果我们在知识库中添加另一句"皮蒂是企鹅"，得出的结论是"皮蒂无法飞"，则上述结论无效。

非单调推理的优点：

对于机器人导航等现实系统，我们可以使用非单调推理。

在非单调推理中，我们可以选择概率事实，也可以做假设。

非单调推理的缺点：

在非单调推理中，可以通过添加新句子来使已有事实无效。

它不能用于定理证明。

Exercises

[Ex. 1] Answer the following questions according to Text A.

1. What are humans best at?

2. What is the kind of knowledge which needs to be represented in AI systems?

3. What is the central component of the knowledge-based agents? What is it represented as?

4. What is declarative knowledge?

5. What is procedural knowledge?

6. What does structural knowledge describe?

7. What components does an artificial intelligence system have for displaying intelligent behavior?

8. What is simple relational knowledge famous for?

9. What does procedural knowledge approach use small programs and codes to do?

10. What properties must a good knowledge representation system possess?

[Ex. 2] Answer the following questions according to Text B.

1. What is reasoning?

2. Why is reasoning essential in artificial intelligence?
3. What is deductive reasoning?
4. What is inductive reasoning?
5. What is inductive reasoning also known as?
6. What is abductive reasoning?
7. What is common sense reasoning?
8. Why is monotonic reasoning not useful for the real-time systems?
9. What are the advantages of monotonic reasoning?
10. When will logic be said as non-monotonic?

[Ex. 3] Translate the following terms or phrases from English into Chinese and vice versa.

1. formal logic   1. _____
2. heuristic knowledge   2. _____
3. imperative knowledge   3. _____
4. declarative knowledge   4. _____
5. structural knowledge   5. _____
6. n.察觉，觉悟，意识   6. _____
7. n.行为   7. _____
8. vt.捕获，捕捉   8. _____
9. n.数据库；资料库；信息库   9. _____
10. n.框架   10. _____

[Ex. 4] Translate the following passages into Chinese.

Knowledge Base

1. What Is a Knowledge Base

A knowledge base is an organized, curated collection of information about a particular subject area. A knowledge base can encompass many forms of content, including:

- Frequently asked questions
- Step-by-step process guides
- Introductory articles
- Video demonstrations
- Glossaries and definition lists

Knowledge bases are the end product of collecting and organizing all of that information into a useful form, through a process called "knowledge management". So typically you would apply knowledge management processes to collecting information, then use knowledge base software to create, manage and deliver that information, as a knowledge base, to your readers.

2. What Is a Knowledge Base Used for

Some types of knowledge bases are intended purely for machines to learn from. Other types are built for people to use and learn from.

Knowledge bases can be aimed at external audiences or internal audiences, and can serve many different purposes. For example, an appliance company may maintain FAQs and maintenance instructions in a customer-facing knowledge base, and also have an internal knowledge base for their employees to understand company policies and learn work related tools.

In some cases, companies maintain knowledge bases that are relevant not just to their own customers, but to anyone interested in their particular field.

3. Why Is a Knowledge Base Important

Your knowledge base isn't just helpful for customers, it's also useful for your staff. A well-structured, clearly written, and cleanly designed knowledge base helps customers help themselves, acts as a learning tool for new staff, and can even be a source for machine learning.

A knowledge base is a cost-effective way to reduce the time and effort the customer has to spend in order to get an answer and move on with their current task.

[Ex. 5] Fill in the blanks with the words given below.

| backward-chaining | conclusion | correctness | inference | containing |
| reasoning | expert | focus | applicable | situation |

1. What Is Rule-Based Reasoning

A particular type of reasoning which uses "if-then-else" rule statements. Rules are simply patterns and an ___1___ engine searches for patterns in the rules that match patterns in the data. The "if" means "when the condition is true," the "then" means "take action A" and the "else" means "when the condition is not true take action B."

Rules can be forward-chaining, also known as data-driven ___2___, because they start with data or facts and look for rules which apply to the facts until a goal is reached. Rules can also be ___3___, also known as goal-driven reasoning, because they start with a goal and look for rules which apply to that goal until a conclusion is reached.

2. What Is an Inference Engine

Software code which processes the rules, cases, objects or other type of knowledge and expertise based on the facts of a given ___4___. Most AI tools contain some form of deductive or inductive reasoning capability.

3. What Is an Expert System

Simply put, an expert system represents information and searches for patterns in that information. They are known as ___5___ systems because they model how a human expert analyzes a particular situation by applying rules to the facts (or compares the current case with similar cases) in order to reach a ___6___. Expert systems can include different types of reasoning like rule-based, case-based, fuzzy logic, neural networks, bayesian networks, etc.

The most common expert system is rule-based, containing a knowledge base (rules) and an inference engine (routing mechanism) which analyzes fact patterns and matches the ___7___ rules. Fact patterns are analyzed until either the goal succeeds or all of the rules are processed and the goal fails.

4. What Is a Knowledge Base

A knowledge base is the representation of expertise, wisdom or rules-of-thumb, often represented by rules ___8___ "if-then-else" conditional statements or cases containing various fact patterns. Knowledge bases may also consist of representative objects (excited utterance) within a sub-class (rules against hearsay) and class (rules of evidence) of information. Knowledge bases typically ___9___ on narrow issues, known as a domain, within a particular fact situation.

5. What Are Validation and Verification

Validation is the process of confirming the ___10___ of a given model or assumptions by using measurable inputs to produce definable outputs, both of which can be confirmed and verified. Verification is the process of confirming that an implemented model works as intended.

Unit 4
Algorithms in Artificial Intelligence

Text A
Search Algorithms in Artificial Intelligence

Artificial intelligence is basically the replication of human intelligence through computer systems or machines. It is done through the process of acquisition of knowledge or information and the addition of rules that is used by information (i.e. learning), and then using these rules to derive conclusions (i.e. reasoning) and then self- correction.

扫码听课文

1. Properties of Search Algorithms

1.1 Completeness
A search algorithm is complete when it returns a solution for any input if at least one solution exists for that particular input.

1.2 Optimality
If the solution deduced by the algorithm is the best solution (i.e. it has the lowest path cost), then that solution is considered as the optimal solution.

1.3 Time and Space Complexity
Time complexity is the time taken by an algorithm to complete its task and space complexity is the maximum storage space needed during the search operation.

2. Types of Search Algorithms

2.1 Uninformed Search Algorithms

Uninformed search algorithms do not have any domain knowledge. They work in a brute force manner, hence they are also called brute force algorithms. They have no knowledge about how far the goal node is. They only know the way to traverse and to distinguish between a leaf node and the goal node. They examine every node without any prior knowledge, hence they are also called blind search algorithms.

Uninformed search algorithms are of mainly two types: Breadth First Search (BFS), and Depth First Search (DFS).

2.1.1 Breadth First Search(BFS)

In breadth first search, the tree or the graph is traversed breadthwise, i.e. it starts from a node called search key and then explores all the neighboring nodes of the search key at that depth first and then moves to the next level nodes. It is implemented using the queue data structure that works on the concept of first in first out (FIFO). It is a complete algorithm as it returns a solution if a solution exists.

The disadvantage of this algorithm is that it requires a lot of memory space because it has to store each level of nodes for the next one. It may also check duplicate nodes.

Some of the applications of BFS are :

- Unweighted Graphs: BFS algorithm can easily create the shortest path and a minimum spanning tree to visit all the vertices of the graph in the shortest time possible with high accuracy.
- P2P Networks: BFS can be implemented to locate all the nearest or neighboring nodes in a peer to peer network. This will find the required data faster.
- Web Crawlers: Search engines or web crawlers can easily build multiple levels of indexes by employing BFS. BFS implementation starts from the source, which is the web page, and then it visits all the links from that source.
- Network Broadcasting: A broadcasted packet is guided by the BFS algorithm to find and reach all the nodes it has the address for.

2.1.2 Depth First Search(DFS)

In depth first search, the tree or the graph is traversed depthwise, i.e. it starts from a node called search key and then explores all the nodes along the branch then backtracks. It is implemented using a stack data structure that works on the concept of last in first out (LIFO).

It stores nodes linearly, hence less space requirement.

The major disadvantage is that this algorithm may go in an infinite loop.

Some of the important applications of DFS are:
- Weighted graph: In a weighted graph, DFS graph traversal generates the shortest path tree and minimum spanning tree.
- Path finding: We can specialize in the DFS algorithm to search a path between two vertices.
- Topological sorting: It is primarily used for scheduling jobs from the given dependencies among the group of jobs. In computer science, it is used in instruction scheduling, data serialization, logic synthesis, determining the order of compilation tasks.
- Searching strongly connected components of a graph: It is used in DFS graph when there is a path from each and every vertex in the graph to other remaining vertexes.
- Solving puzzles with only one solution: DFS algorithm can be easily adapted to search all solutions to a maze by including nodes on the existing path in the visited set.

The differences between DFS and BFS are:
- BFS finds the shortest path to the destination whereas DFS goes to the bottom of a subtree, then backtracks.
- BFS uses a queue to keep track of the next location to visit. whereas DFS uses a stack to keep track of the next location to visit.
- BFS traverses according to tree level while DFS traverses according to tree depth.
- BFS is implemented using FIFO list while DFS is implemented using LIFO list.
- In BFS, you can never be trapped into finite loops whereas in DFS you can be trapped into infinite loops.

2.2 Informed Search Algorithms

Informed search algorithms have domain knowledge. They contain the problem description as well as extra information like how far the goal node is. They are also called the heuristic search algorithm. They might not give the optimal solution always but they will definitely give a good solution in a reasonable time. They can solve complex problems more easily than uninformed.

They are mainly of two types: greedy best first search, and A* search.

2.2.1 Greedy Best First Search

In this algorithm, we expand the closest node to the goal node. The closeness factor is roughly calculated by heuristic function h(x). The node is expanded or explored when f(n)=h(n). This algorithm is implemented through the priority queue. It is not an optimal algorithm. It can get stuck in loops.

2.2.2 A* Search

A* search is a combination of greedy search and uniform cost search. In this algorithm, the total cost (heuristic) which is denoted by f(x) is a sum of the cost in uniform cost search denoted by g(x) and cost of greedy search denoted by h(x).

$$f(x) = g(x) + h(x)$$

In this g(x) is the backward cost which is the cumulative cost from the root node to the current node and h(x) is the forward cost which is approximate of the distance of goal node and the current node.

3. Conclusion

In this article, various artificial intelligence search algorithms are explained. AI is growing at a rapid rate and it is acquiring the market and search algorithms are an important part of artificial intelligence.

New Words

replication	[ˌrepli'keiʃn]	n. 复制；折叠
acquisition	[ˌækwi'ziʃn]	n. 获得
derive	[di'raiv]	v. 得到，导出
self-correction	['self kə'rekʃn]	v. 自我纠错，自我改正，自校正
completeness	[kəm'pli:tnis]	n. 完全，完全性，完整性
particular	[pə'tikjʊlə]	adj. 特别的；详细的；独有的
optimality	[ˌɒpti'mæləti]	n. 最优性；最佳性
path	[pɑ:θ]	n. 路径
optimal	['ɒptiməl]	adj. 最佳的，最优的
complexity	[kəm'pleksəti]	n. 复杂性，复杂度
uninformed	[ˌʌnin'fɔ:md]	adj. 信息不足的；情况不明的
prior	['praiə]	adj. 优先的，占先的；在……之前
breadthwise	['bredθwaiz]	adv. 横向地，在广度方向上
queue	[kju:]	n. 队列
memory	['meməri]	n. 存储器，内存
duplicate	['dju:plikeit]	v. 重复；复制
	['dju:plikit]	adj. 重复的；复制的；副本的
spanning	['spæniŋ]	adj.（尤指树形子图）生成的
vertices	['vɜ:rtisi:z]	n. 最高点（vertex 的名词复数）；顶点

graph	[græf]	n.图表，曲线图
crawler	['krɔ:lə]	n.爬虫
index	['indeks]	n.索引；<数>指数
link	[liŋk]	n.链接
broadcast	['brɔ:dkɑ:st]	vt.广播；播放
packet	['pækit]	n.包；信息包
depthwise	[depθwaiz]	adv.纵向地，在深度方向上
stack	[stæk]	n.堆栈
linearly	['liniəli]	adv.线性地
topological	[tɒpə'lɒdʒikəl]	adj.拓扑的
synthesis	['sinθisis]	n.综合
maze	[meiz]	n.迷宫
subtree	['sʌbtri:]	n.子树
backtrack	['bæktræk]	vi.回溯，由原路返回
finite	['fainait]	adj.有限的
loop	[lu:p]	n.循环
infinite	['infinət]	adj.无限的，无穷的
extra	['ekstrə]	adj.额外的，补充的
reasonable	['ri:znəbl]	adj.合理的，适当的
greedy	['gri:di]	adj.贪婪的，贪心的
factor	['fæktə]	n.因素
function	['fʌŋkʃn]	n.函数
priority	[prai'ɒrəti]	n.优先，优先权
cumulative	['kju:mjʊlətiv]	adj.累积的；渐增的；追加的

Phrases

computer system	计算机系统
search algorithm	搜索算法
path cost	路径成本，路径代价
be considered as	被认为是
optimal solution	最优解
time complexity	时间复杂度

space complexity	空间复杂度
storage space	存储空间
uninformed search algorithm	不知情搜索算法
brute force	蛮力，强力
leaf node	叶节点
blind search algorithm	盲目搜索算法
neighboring node	相邻节点
data structure	数据结构
a lot of	许多的，大量的
unweighted graph	无权重图
minimum spanning tree	最小生成树
web crawler	网页爬虫，网络爬虫
search engine	搜寻引擎
web page	网页
shortest path tree	最短路径树
data serialization	数据序列化
keep track of	跟踪，记录
be trapped into	陷入，被困在
informed search algorithm	知情搜索算法
heuristic search algorithm	启发式搜索算法
greedy best first search	贪婪最佳优先搜索
be denoted by ...	由……表示
backward cost	向后成本
forward cost	向前成本

Abbreviations

BFS (Breadth First Search)	广度优先搜索
DFS (Depth First Search)	深度优先搜索
FIFO (First In First Out)	先进先出
P2P (Peer to Peer)	对等网络
LIFO (Last In First Out)	后进先出

Text A 参考译文
人工智能中的搜索算法

人工智能基本上是通过计算机系统或机器对人类智能的复制。它是通过获取知识或信息并添加信息使用的规则（即学习）的过程来完成的，然后使用这些规则来得出结论（即推理），之后进行自我校正。

1. 搜索算法的性质

1.1 完整性
如果针对特定输入存在至少一个解决方案，则该搜索算法返回针对任何输入的解决方案时搜索即告完成。

1.2 最优性
如果算法推导的解决方案是最佳解决方案（即路径成本最低），则将该解决方案视为最佳解决方案。

1.3 时空复杂度
时间复杂度是算法完成其任务所需的时间，而空间复杂度是搜索操作期间所需的最大存储空间。

2. 搜索算法的类型

2.1 不知情搜索算法
不知情搜索算法没有任何域知识。它以蛮力方式工作，因此也称为蛮力算法。它不了解目标节点有多远，只知道遍历和区分叶节点和目标节点的方式。它在没有任何先验知识的情况下检查每个节点，因此也称为盲搜索算法。

不知情搜索算法主要分为两种：广度优先搜索（BFS）和深度优先搜索（DFS）。

2.1.1 广度优先搜索（BFS）
在广度优先搜索中，树或图在广度方向上遍历，即，它从称为搜索键的节点开始，然后首先根据广度优先来搜索该键的所有相邻节点，之后移至下一级节点。它是使用队列数据结构实现的，该数据结构采用先进先出（FIFO）的概念。这是一个完整的算法，因为如果存在解决方案，它将返回一个解决方案。

该算法的缺点是它需要大量的存储空间，因为它必须存储下一节点的每个级别的节点。它还可能检查重复的节点。

BFS 有以下一些应用：
- 非加权图：BFS 算法可以轻松创建最短路径和最小生成树，从而在最短时间内以高准确性访问图的所有顶点。
- P2P 网络：可以实施 BFS 来定位对等网络中所有最近或相邻的节点，这样可以更快地找到所需的数据。
- 网页爬虫程序：搜索引擎或网页爬虫程序可以通过使用 BFS 轻松构建多个级别的索引。BFS 实施从源（即网页）开始，然后访问该源的所有链接。
- 网络广播：广播的数据包由 BFS 算法引导，以查找并到达其地址所在的所有节点。

2.1.2 深度优先搜索（DFS）

在深度优先搜索中，树或图在深度方向上遍历，即，它从称为搜索键的节点开始，然后沿分支探索所有节点，然后回溯。它是使用堆栈数据结构实现的，该结构适用于后进先出（LIFO）的概念。

它线性地存储节点，因此需要的空间更少。

该算法的主要缺点是可能会陷入无限循环。

DFS 有以下一些重要应用：
- 加权图：在加权图中，DFS 图遍历会生成最短路径树和最小生成树。
- 路径查找：我们可以专门研究 DFS 算法，以搜索两个顶点之间的路径。
- 拓扑排序：主要用于根据作业组中给定的依存关系调度作业。在计算机科学中，它用于指令调度、数据序列化、逻辑综合、确定编译任务的顺序。
- 搜索图的强连接组件：当从图中的每个顶点到其他顶点都有路径时，它在 DFS 图中使用。
- 仅用一种解决方案即可解决难题：通过包含已访问集合中现有路径上的节点，可以轻松地找到迷宫的所有解决方案。

DFS 和 BFS 的区别：
- BFS 找到抵达目标的最短路径，而 DFS 到达子树的底部，然后回溯。
- BFS 使用队列来跟踪下一个要访问的位置，而 DFS 使用堆栈来跟踪下一个要访问的位置。
- BFS 根据树层级遍历，而 DFS 根据树深度遍历。
- BFS 使用 FIFO 列表实现，而 DFS 使用 LIFO 列表实现。
- 在 BFS 中，永远不会陷入有限循环，而在 DFS 中，可能会陷入无限循环。

2.2 知情搜索算法

知情搜索算法具有域知识，它包含问题描述以及其他信息，例如目标节点的距离，也被称为启发式搜索算法。它可能不会始终提供最佳解决方案，但肯定会在合理的时间内提供良好的解决方案。它比不知情搜索算法更容易能解决复杂的问题。

它主要有两种类型：贪婪最佳优先搜索和 A*搜索。

2.2.1 贪婪最佳优先搜索

在该算法中，我们将最接近的节点扩展到目标节点。通过启发函数 $h(x)$ 粗略地计算出接近度因子。当 $f(n)=h(n)$ 时，节点被展开或探索。该算法是通过优先级队列实现的，这不是最佳算法，它可能卡在循环中。

2.2.2 A*搜索

A*搜索是贪婪搜索和统一成本搜索的组合。在该算法中，$f(x)$ 表示的总成本（启发式）是 $g(x)$ 表示的统一成本搜索的成本与由 $h(x)$ 表示的贪婪搜索成本之和。

$$f(x)=g(x)+h(x)$$

$g(x)$ 是后向成本，它是从根节点到当前节点的累积成本；$h(x)$ 是前向成本，它近似于目标节点与当前节点的距离。

3. 结论

本文介绍了各种人工智能搜索算法。人工智能正快速增长，它正在占领市场，搜索算法是人工智能的重要组成部分。

Text B
Machine Learning Algorithms

扫码听课文

The term "machine learning" is often, incorrectly, interchanged with artificial intelligence. Machine learning is actually a sub field or type of AI. Machine learning is also often referred to as predictive analytics, or predictive modelling.

Coined by American computer scientist Arthur Samuel in 1959, the term "machine learning" is defined as a "computer's ability to learn without being explicitly programmed".

At its most basic, machine learning uses programmed algorithms that receive and analyse input data to predict output values within an acceptable range. As new data is fed to these algorithms, they learn and optimise their operations to improve performance, developing "intelligence" over time.

There are four types of machine learning algorithms: supervised, semi-supervised, unsupervised and reinforcement.

1. Supervised Learning

In supervised learning, the machine is taught by example. The operator provides the machine learning algorithm with a known data set that includes desired inputs and outputs, and the algorithm must find a method to determine how to arrive at those inputs and outputs. While the operator knows the correct answers to the problem, the algorithm identifies patterns in data, learns from observations and makes predictions. The algorithm makes predictions and is corrected by the operator and this process continues until the algorithm achieves a high level of accuracy/performance.

Under the umbrella of supervised learning fall: classification, regression and forecasting.

Classification: In classification tasks, the machine learning program must draw a conclusion from observed values and determine what category new observations belong to. For example, when filtering emails as "spam" or "not spam", the program must look at existing observational data and filter the emails accordingly.

Regression: In regression tasks, the machine learning program must estimate and understand the relationships among variables. Regression analysis focuses on one dependent variable and a series of other changing variables, making it particularly useful for prediction and forecasting.

Forecasting: Forecasting is the process of making predictions about the future based on the past and present data, and is commonly used to analyse trends.

2. Semi-Supervised Learning

Semi-supervised learning is similar to supervised learning, but it uses both labelled and unlabelled data. Labelled data is essentially information that has meaningful tags so that the algorithm can understand the data, whilst unlabelled data lacks that information. By using this combination, machine learning algorithms can learn to label unlabelled data.

3. Unsupervised Learning

Here, the machine learning algorithm studies data to identify patterns. There is no answer key or human operator to provide instruction. Instead, the machine determines the correlations and relationships by analysing available data. In an unsupervised learning process, the machine learning algorithm is left to interpret large data sets and address that data accordingly. The algorithm tries to organise that data in some way to describe its structure. This might mean grouping the data into clusters or arranging it in a way that looks more organised.

As it assesses more data, its ability to make decisions on that data gradually improves and becomes more refined.

Under the umbrella of unsupervised learning, fall:

Clustering: Clustering involves grouping sets of similar data (based on defined criteria). It's useful for segmenting data into several groups and performing analysis on each data set to find patterns.

Dimension reduction: Dimension reduction reduces the number of variables being considered to find the exact information required.

4. Reinforcement Learning

Reinforcement learning focuses on regimented learning processes, where a machine learning algorithm is provided with a set of actions, parameters and end values. By defining the rules, the machine learning algorithm tries to explore different options and possibilities, monitor and evaluate each result to determine which one is optimal. Reinforcement learning teaches the machine trial and error. It learns from past experiences and begins to adapt its approach in response to the situation to achieve the best possible result.

5. What Machine Learning Algorithms Can You Use

Choosing the right machine learning algorithm depends on several factors, including, but not limited to: Data size, quality and diversity, as well as what answers businesses want to derive from that data. Additional considerations include accuracy, training time, parameters, data points and much more. Therefore, choosing the right algorithm is both a combination of business need, specification, experimentation and time available. Even the most experienced data scientists cannot tell you which algorithm will perform the best before experimenting with others. We have, however, compiled a machine learning algorithm "cheat sheet" which will help you find the most appropriate one for your specific challenges.

6. What Are the Most Common and Popular Machine Learning Algorithms

6.1 Naive Bayes Classifier Algorithm (Supervised Learning - Classification)

The naive Bayes classifier is based on Bayes' theorem and classifies every value as independent of any other value. It allows us to predict a class/category based on a given set of features using probability.

Despite its simplicity, the classifier does surprisingly well and is often used due to the fact it outperforms more sophisticated classification methods.

6.2 K Means Clustering Algorithm (Unsupervised Learning - Clustering)

The K means clustering algorithm is a type of unsupervised learning, which is used to categorise unlabelled data, i.e. data without defined categories or groups. The algorithm works by finding groups within the data, with the number of groups represented by the variable K. It then works iteratively to assign each data point to one of K groups based on the features provided.

6.3 Support Vector Machine Algorithm (Supervised Learning - Classification)

Support vector machine algorithms are supervised learning models that analyse data used for classification and regression analysis. They essentially filter data into categories, which is achieved by providing a set of training examples, each set marked as belonging to one or the other of the two categories. The algorithm then works to build a model that assigns new values to one category or the other.

6.4 Linear Regression (Supervised Learning/Regression)

Linear regression is the most basic type of regression. Simple linear regression allows us to understand the relationships between two continuous variables.

6.5 Logistic Regression (Supervised learning – Classification)

Logistic regression focuses on estimating the probability of an event occurring based on the previous data provided. It is used to cover a binary dependent variable, that is where only two values, 0 and 1, represent outcomes.

6.6 Artificial Neural Networks (Reinforcement Learning)

An artificial neural network (ANN) comprises "units" arranged in a series of layers, each of which connects to layers on either side. ANNs are inspired by biological systems, such as the brain, and how they process information. ANNs are essentially a large number of interconnected processing elements working in unison to solve specific problems.

ANNs also learn by example and through experience, and they are extremely useful for modelling non-linear relationships in high-dimensional data or where the relationship among the input variables is difficult to understand.

6.7 Decision Trees (Supervised Learning – Classification/Regression)

A decision tree is a flow-chart-like tree structure that uses a branching method to illustrate every possible outcome of a decision. Each node within the tree represents a test on a specific variable and each branch is the outcome of that test.

6.8 Random Forests (Supervised Learning – Classification/Regression)

Random forests or "random decision forests" is an ensemble learning method, combining multiple algorithms to generate better results for classification, regression and other tasks. Each individual classifier is weak, but when combined with others, it can produce excellent results. The algorithm starts with a "decision tree" (a tree-like graph or model of decisions) and an input is entered at the top. It then travels down the tree, with data being segmented into smaller and smaller sets, based on specific variables.

6.9 K-Nearest Neighbour Algorithm (Supervised Learning)

The K-nearest neighbour algorithm estimates how likely a data point is to be a member of one group or another. It essentially looks at the data points around a single data point to determine what group it is actually in. For example, if one data point is on a grid, the algorithm will try to determine what group that data point is in (Group A or Group B, for example), and it will look at the data points near it to see what group the majority of the points are in.

New Words

predictive	[prɪˈdɪktɪv]	adj. 预言性的
explicitly	[ɪkˈsplɪsɪtli]	adv. 明白地，明确地
range	[reɪndʒ]	n. 范围
optimise	[ˈɒptɪmaɪz]	vt. 使最优化
supervised	[ˈsuːpəvaɪzd]	adj. 监督的
semi-supervised	[ˈsemi ˈsuːpəvaɪzd]	adj. 半监督的
unsupervised	[ˌʌnsjuːˈpəvaɪzd]	adj. 无监督的
operator	[ˈɒpəreɪtə]	n. 操作员；运算符
regression	[rɪˈɡreʃn]	n. 回归
forecasting	[ˈfɔːkɑːstɪŋ]	n. 预测
conclusion	[kənˈkluːʒn]	n. 结论；断定；推论
filter	[ˈfɪltə]	v. 过滤，滤除
email	[ˈiːmeɪl]	n. 电子邮件

spam	[spæm]	n.垃圾邮件
observational	[ˌɒbzə'veiʃənəl]	adj.观察的，观测的
label	['leibl]	n.标记；标签
		vt.加标记于，贴标签于
organise	['ɔ:gənaiz]	v.组织；安排
structure	['strʌktʃə]	n.结构；构造；体系
cluster	['klʌstə]	n.丛；簇，串；群
		vi.丛生；群聚
		vt.使密集，使聚集
gradually	['grædʒʊəli]	adv.逐步地，渐渐地
refine	[ri'fain]	vt.提炼；改善
criteria	[krai'tiəriə]	n.标准，准则
dimension	[dai'menʃn]	n.度，维；尺寸
regimented	['redʒiməntid]	adj.有条理的；严格规划的
monitor	['mɒnitə]	v.监视；控制
evaluate	[i'væljʊeit]	v.评价，估价
trial	['traiəl]	n.试验
response	[ri'spɒns]	n.反应，响应；回答
diversity	[dai'vɜ:səti]	n.多样化，多样性
specification	[ˌspesifi'keiʃn]	n.规范，规格；说明书
experimentation	[ikˌsperimen'teiʃn]	n.实验，试验
naive	[nai'i:v]	adj.朴素的；单纯的
probability	[ˌprɒbə'biləti]	n.概率，可能性
simplicity	[sim'plisəti]	n.简单，朴素
surprisingly	[sə'praiziŋli]	adv.惊人地，出人意外地
outperform	[ˌaʊtpə'fɔ:m]	vt.做得比……更好，胜过
variable	['veəriəbl]	adj.变量的
		n.变量，可变因素
iteratively	['itərətivli]	adv.迭代地
continuous	[kən'tinjʊəs]	adj.连续的
comprise	[kəm'praiz]	vt.包含，包括；由……组成；由……构成
layer	['leiə]	n.层
connect	[kə'nekt]	vi.连接，建立关系
inspire	[in'spaiə]	vt.启发，启迪

element	['elimənt]	n. 元素；要素
random	['rændəm]	adj. 随机的，任意的
grid	[grid]	n. 栅格，格子
majority	[mə'dʒɒrəti]	n. 多数

Phrases

interchange with...	与……互换
sub field	子域
predictive modelling	预测建模
semi-supervised learning	半监督学习
dimension reduction	降维
reinforcement learning	强化学习
derive from	由……起源；取自
cheat sheet	备忘单
naive Bayes classifier algorithm	朴素贝叶斯分类器算法
K means clustering algorithm	K 均值聚类算法
support vector machine algorithm	支持向量机算法
regression analysis	回归分析
linear regression	线性回归
logistic regression	逻辑回归
in union	齐心协力，整齐划一
decision tree	决策树
flow-chart-like tree	流程图状的树
be segmented into	被分割成
K-nearest neighbour algorithm	K 最近邻算法

Text B 参考译文
机器学习算法

人们通常错误地把术语"机器学习"与人工智能等同。机器学习实际上是人工智能的一个子领域或类型，通常也称为预测分析或预测建模。

"机器学习"一词由美国计算机科学家亚瑟·塞缪尔于1959年提出,其定义是"计算机无须明确编程即可学习的能力"。

机器学习最基本的方法是使用接收和分析输入数据的编程算法来预测可接受范围内的输出值。随着将新数据馈入这些算法,他们将学习并优化其操作以提高性能,并逐步发展"智能"。

机器学习算法有四种类型:监督学习算法、半监督学习算法、无监督学习算法和强化学习算法。

1. 监督学习

在监督学习中,机器依例而学。操作员向机器学习算法提供一个包含所需输入和输出的已知数据集,并且该算法必须找到一种方法来确定如何得出这些输入和输出。操作员知道问题的正确答案,该算法可识别数据中的模式,从观察中学习并做出预测。该算法进行预测并由操作员进行校正,并且该过程一直持续到该算法达到较高的准确性/性能水平为止。

监督学习可以分为分类、回归和预测。

分类:在分类任务中,机器学习程序必须从观测值得出结论,并确定新观测值属于哪一类别。例如,当将电子邮件过滤为"垃圾邮件"或"非垃圾邮件"时,程序必须查看现有的观察数据并相应地过滤电子邮件。

回归:在回归任务中,机器学习程序必须估计并了解变量之间的关系。回归分析着重于一个因变量和一系列其他不断变化的变量,这使其对预报和预测特别有用。

预测:预测是根据过去和现在的数据对未来进行预测的过程,通常用于分析趋势。

2. 半监督学习

半监督学习与监督学习相似,但是它同时使用了标记和未标记的数据。标记数据本质上是具有有意义标签的信息,因此算法可以理解该数据,而未标记数据则缺少该信息。通过使用此组合,机器学习算法可以学习标记那些未标记的数据。

3. 无监督学习

在这里,通过机器学习算法研究数据来识别模式,此处没有参考答案,也没有人工操作员提供指导。相反的是,由机器通过分析现有数据来确定相关性和关系。在无监督的学习过程中,机器学习算法被用来解释大型数据集并相应地处理该数据。该算法尝试以某种方式组织数据以描述其结构,这可能意味着将数据分组到聚类中或以看起来更有条理的方式进行整理。

当它评估更多数据时,其对该数据进行决策的能力逐渐提高并更加完善。

无监督学习分为:

聚类:聚类是将相似数据集(基于定义的标准)进行分组,以便于将数据分成几组并对每组数据进行分析以找到模式。

降维：要找到所需的确切信息需要考虑一些变量，而降维减少了这些变量的数量。

4. 强化学习

强化学习专注于有条理的学习过程，其中提供了有一组动作、参数和最终值的机器学习算法。通过定义规则，机器学习算法尝试探索不同的选择和可能性，监视和评估每个结果以确定哪个是最佳选择。强化学习可指导机器反复试验，它从过去的经验中汲取教训，并开始根据情况调整其方法以达成最佳结果。

5. 你可以使用哪些机器学习算法

选择正确的机器学习算法取决于几个因素，包括但不限于：数据大小、质量和多样性以及企业希望从该数据中得出什么答案。其他注意事项包括准确性、训练时间、参数、数据点等。因此，选择正确的算法是业务需求、规范、实验和可用时间的结合。即使是经验最丰富的数据科学家，也无法在试验其他算法之前就能告诉你哪种算法性能最好。但是，我们已经编译了一种机器学习算法"备忘单"，可以帮助你找到最适合特定挑战的算法。

6. 最常见和最受欢迎的机器学习算法是什么

6.1 朴素贝叶斯分类器算法（监督学习-分类）

朴素贝叶斯分类器基于贝叶斯定理，将每个值独立于其他任何值进行分类。它使我们能够使用概率基于给定的一组特征来预测分类/类别。

尽管分类器简单，但其表现却出奇地好，并且由于其胜过更复杂的分类方法而经常被使用。

6.2 K均值聚类算法（无监督学习-聚类）

K均值聚类算法是一种无监督学习，用于对未标记的数据（即没有定义类别或组的数据）进行分类。该算法的工作原理是在数据中查找组，并用变量 K 表示组的数量，然后，根据提供的功能，迭代地将每个数据点分配给一个 K 组。

6.3 支持向量机算法（监督学习-分类）

支持向量机算法是监督学习模型，可以分析用于分类和回归分析的那些数据。它们实质上是将数据过滤到类别中，这可以通过提供一组训练示例来实现。每组训练示例都标记为属于两个类别中的一个或另一个，然后，该算法将构建一个将新值分配给一个类别或另一个类别的模型。

6.4 线性回归（监督学习/回归）

线性回归是最基本的回归类型。简单的线性回归使我们能够理解两个连续变量之间的关系。

6.5 逻辑回归（监督学习-分类）

逻辑回归专注于根据先前提供的数据来估计事件发生的可能性，它用于覆盖二进制因变量，即只用两个值（0和1）表示结果。

6.6 人工神经网络（强化学习）

人工神经网络（ANN）包含布置在一系列层中的"单元"，每个单元都连接到任一侧的层。人工神经网络受到诸如大脑之类的生物系统以及它们如何处理信息的启发。人工神经网络本质上是大量相互连接的处理元素，它们协同工作以解决特定问题。

人工神经网络还可以通过实例和经验学习，它们对于在高维数据中建立非线性关系模型非常有用，在输入变量之间的关系难以理解时也非常有用。

6.7 决策树（监督学习 – 分类/回归）

决策树是一种类似于流程图的树结构，它使用分支方法来说明决策的每种可能结果。树中的每个节点代表一个针对特定变量的测试，每个分支都是该测试的结果。

6.8 随机森林（监督学习 – 分类/回归）

随机森林或"随机决策森林"是一种整体学习方法，结合了多种算法，可以为分类、回归和其他任务生成更好的结果。每个单独的分类器都很弱，但是与其他分类器结合使用时，可以产生出色的结果。该算法以"决策树"（树状图或决策模型）开头，并在顶部输入，然后，它沿着树行进，并根据特定变量将数据分割成越来越小的集合。

6.9 K最近邻算法（监督学习）

K最近邻算法估计数据点成为一个或另一个组的成员的可能性。它实质上是查看单个数据点周围的数据点，以确定它实际位于哪个组中。例如，如果一个数据点在网格上，该算法会试图确定该数据点在哪个组中（例如A组或B组），它会查看该数据点周围的数据点，看看大多数点都在哪个组。

Exercises

[Ex. 1] Answer the following questions according to Text A.

1. When is a search algorithm complete?
2. When is that solution considered as the optimal solution?
3. What is time complexity?

4. Why are uninformed search algorithms also called brute force algorithms and blind search algorithms?

5. In breadth first search, how is the tree or the graph traversed? How is breadth first search implemented?

6. In depth first search, how is the tree or the graph traversed? How is depth first search implemented?

7. What is the disadvantage of breadth first search algorithm? Why? And what is the major disadvantage of depth first search algorithm?

8. What are some of the important applications of DFS?

9. Do informed search algorithms have domain knowledge? What do they contain?

10. What are the two main types of informed search algorithms?

[Ex. 2] Answer the following questions according to Text B.

1. What is machine learning actually? What is it also referred to? Who coined it and when?

2. How many types of machine learning algorithms are there? What are they?

3. How is the machine taught in supervised learning? What does the operator provide the machine with? And what must the algorithm do?

4. What fall under the umbrella of supervised learning?

5. What is the difference between labelled data and unlabelled data?

6. What is the machine learning algorithm left to do in an unsupervised learning process?

7. What fall under the umbrella of unsupervised learning?

8. What does reinforcement learning focus on?

9. What is the K means clustering algorithm? How does it work?

10. What is a decision tree?

[Ex. 3] Translate the following terms or phrases from English into Chinese and vice versa.

1.	data structure	1.	
2.	greedy best first search	2.	
3.	heuristic search algorithm	3.	
4.	informed search algorithm	4.	
5.	shortest path tree	5.	
6.	vi.回溯，由原路返回	6.	
7.	n.复杂性，复杂度	7.	

8. *n.*完全，完全性，完整性 8. _____
9. *adj.*有限的 9. _____
10. *n.*函数 10. _____

[Ex. 4] Translate the following passages into Chinese.

Categories of Artificial Intelligence Algorithm

Artificial intelligence algorithm is a broad field which consists of machine learning algorithms as well as deep learning algorithms. The main goal of these algorithms is to enable computers to learn on their own and make a decision or find useful patterns. Artificial intelligence algorithms learn from the data itself. In a broader sense learning can be divided into 3 categories:

• Supervised learning: When the labels of both input and output are known and the model learns from data to predict output for similar input data, it is supervised learning.

• Unsupervised learning: When output data is unknown or it is needed to find patterns in data given, such type of learning is unsupervised learning.

• Reinforcement learning: Algorithms learns to perform an action from experience. Here algorithms learn through trial and error which action yields greatest rewards. The objective is to choose actions that maximize the expected reward over a given amount of time.

According to problems which humans encounter and they solve, there are three categories in which these algorithms can be divided to perform the same actions.

• Classification: Humans do make decision based on classification, for example, will this shirt look good on me or not? Here the human mind will process some algorithm with previous experience (data) and then the output will be yes or no. In the same way, these classification algorithms will take some input data and based on this it will predict yes or no. Some examples of these algorithms are Naïve Bayes, logistic regression, SVM, etc.

• Regression: Here the output is continuous and there is no specific category. For example, what will be the temperature tomorrow? A human mind will think of season and temperature of previous days and will predict some number. Some examples of these algorithms are linear regression, gradient boosting, random forest, etc.

• Clustering: Sometimes we don't have to make a decision on given input but to distinguish odd ones. For example, seeing a painting and finding different patterns. Some examples of these algorithms are K-means clustering, hierarchical clustering, etc.

[Ex. 5] Fill in the blanks with the words given below.

value	neighbouring	maximize	distance	solutions
limited	condition	state	optimizing	computation

Hill Climbing Algorithm

1. Introduction

Hill climbing is a form of heuristic search algorithm which is used in solving optimization related problems in Artificial Intelligence domain. The algorithm starts with a non-optimal __1__ and iteratively improves its state until some predefined condition is met. The __2__ to be met is based on the heuristic function. The aim of the algorithm is to reach an optimal state which is better than its current state. The starting point which is the non-optimal state is referred to as the base of the hill and it tries to constantly iterate (climb) untill it reaches the peak __3__, that is why it is called hill climbing algorithm.

Hill climbing algorithm is a memory-efficient way of solving large computational problems. It takes into account the current state and immediate __4__ state. The hill climbing problem is particularly useful when we want to __5__ or minimize any particular function based on the input which it is taking. The most commonly used hill climbing algorithm is "travelling salesman problem" where we have to minimize the __6__ travelled by the salesman. Hill climbing algorithm may not find the global optimal (best possible) solution but it is good for finding local minima/maxima efficiently.

2. Key Features of Hill Climbing

There are few of the key features of hill climbing algorithm as follows:
- Greedy approach: The algorithm moves in the direction of __7__ the cost i.e. finding Local Maxima/Minima.
- No backtracking: It cannot remember the previous state of the system so backtracking to the previous state is not possible.
- Feedback mechanism: The feedback from the previous __8__ helps in deciding the next course of action i.e. whether to move up or down the slope.

3. Advantages of Hill Climbing Algorithm

Advantages of hill climbing algorithm in artificial intelligence are given below:
- Hill climbing is very useful in routing-related problems like travelling salesmen problem, job

scheduling, chip designing and portfolio management.
- It is good in solving the optimization problem while using only __9__ computation power.
- It is more efficient than other search algorithms.

Hill climbing algorithm is a very widely used algorithm for optimization-related problems as it gives decent __10__ to computationally challenging problems. It has certain drawbacks associated with it like its Local Minima, Ridge and Plateau problem which can be solved by using some advanced algorithm.

Unit 5
Expert System and Fuzzy Logic

Text A
Expert System

扫码听课文

An expert system (ES) is a computer program that is designed to solve complex problems and to provide decision-making ability like a human expert. It performs this by extracting knowledge from its knowledge base using the reasoning and inference rules according to the user queries.

The expert system is a part of AI, and the first ES was developed in the year 1970, which was the first successful approach of artificial intelligence. It solves the most complex issue as an expert by extracting the knowledge stored in its knowledge base. The system helps in decision making for complex problems using both facts and heuristics like a human expert. It is called so because it contains the expert knowledge of a specific domain and can solve any complex problem of that particular domain. These systems are designed for a specific domain, such as medicine, science, etc.

The performance of an expert system is based on the expert's knowledge stored in its knowledge base. The more knowledge stored in the KB, the more that system improves its performance. One of the common examples of an ES is a suggestion of spelling errors while typing in the Google search box.

1. Characteristics of Expert System

High performance: The expert system provides high performance for solving any type of complex problem of a specific domain with high efficiency and accuracy.

Understandable: It responds in a way that can be easily understood by the user. It can take input in human language and provides the output in the same way.

Reliable: It is much reliable for generating an efficient and accurate output.

Highly responsive: ES provides the result for any complex query within a very short period of time.

2. Components of Expert System

2.1 User Interface

With the help of a user interface, the expert system interacts with the user, takes queries as an input in a readable format, and passes it to the inference engine. After getting the response from the inference engine, it displays the output to the user. In other words, it is an interface that helps a non-expert user to communicate with the expert system to find a solution.

2.2 Inference Engine (Rules of Engine)

The inference engine is known as the brain of the expert system as it is the main processing unit of the system. It applies inference rules to the knowledge base to derive a conclusion or deduce new information. It helps in deriving an error-free solution of queries asked by the user.

With the help of an inference engine, the system extracts the knowledge from the knowledge base.

There are two types of inference engine.

Deterministic inference engine: The conclusions drawn from this type of inference engine are assumed to be true. It is based on facts and rules.

Probabilistic inference engine: This type of inference engine contains uncertainty in conclusions, and based on the probability.

Inference engine uses the following modes to derive the solutions.

Forward chaining: It starts from the known facts and rules, and applies the inference rules to add their conclusion to the known facts.

Backward chaining: It is a backward reasoning method that starts from the goal and works backward to prove the known facts.

2.3 Knowledge Base

Knowledge base is a type of storage that stores knowledge acquired from different experts of the particular domain. It is considered as big storage of knowledge. The bigger the knowledge base, the more precise the expert system will be.

Knowledge base is similar to a database that contains information and rules of a particular domain or subject. One can also view the knowledge base as collections of objects and their attributes. Such as a Lion is an object and its attributes are it is a mammal, it is not a domestic animal, etc.

The following are the components of knowledge base.

Factual knowledge: The knowledge which is based on facts and accepted by knowledge engineers comes under factual knowledge.

Heuristic knowledge: This knowledge is based on practice, the ability to guess, evaluation, and experiences.

Knowledge representation: It is used to formalize the knowledge stored in the knowledge base using the if-else rules.

Knowledge acquisitions: It is the process of extracting, organizing, and structuring the domain knowledge, specifying the rules for acquiring the knowledge from various experts, and storing that knowledge into the knowledge base.

2.4 Development of Expert System

Here, we will explain the working of an expert system by taking an example of MYCIN ES. Below are some steps to build an MYCIN.

Firstly, ES should be fed with expert knowledge. In the case of MYCIN, human experts specialized in the medical field of bacterial infection provide information about the causes, symptoms, and other knowledge in that domain.

Secondly, the KB of the MYCIN is updated successfully. In order to test it, the doctor provides a new problem to it. The problem is to identify the presence of the bacteria by inputting the details of a patient, including the symptoms, current condition, and medical history.

Thirdly, the ES will need a questionnaire to be filled by the patient to know the general information about the patient, such as gender, age, etc.

Now the system has collected all the information, so it will find the solution for the problem by applying if-then rules using the inference engine and using the facts stored within the KB.

In the end, it will provide a response to the patient by using the user interface.

2.5 Participants in the Development of Expert System

There are three primary participants in the building of expert system.

Expert: The success of an ES depends much on the knowledge provided by human experts. These experts are those persons who are specialized in that specific domain.

Knowledge engineer: Knowledge engineer is the person who gathers the knowledge from the domain experts and then codifies that knowledge to the system according to the formalism.

End-user: This is a particular person or a group of people who may not be experts but working on the expert system and need the solution or advice for his queries, which are complex.

3. Why Expert System

Before using any technology, we must have an idea about why to use that technology and hence the same for the ES. The following are the points that are describing the need of the ES.

No memory limitations: It can store as much data as required and can memorize it at the time of its application. But for human experts, it is very difficult to memorize all things at every time.

High efficiency: If the knowledge base is updated with the correct knowledge, it will provide a highly efficient output, which may not be possible for a human.

Expertise in a domain: There are lots of human experts in each domain, and they all have different skills and different experiences, so it is not easy to get a final output for the query. But if we put the knowledge gained from human experts into the expert system, it will provide an efficient output by mixing all the facts and knowledge.

Not affected by emotions: These systems are not affected by human emotions such as fatigue, anger, depression, anxiety, etc. Hence the performance remains constant.

High security: These systems provide high security to resolve any query.

It considers all facts: To respond to any query, it checks and considers all the available facts and provides the result accordingly. But it is possible that a human expert may not consider some facts due to any reason.

Regular updates improve the performance: If there is an issue in the result provided by the expert systems, we can improve the performance of the system by updating the knowledge base.

4. Capabilities of Expert System

Some capabilities of an expert system are:

Advising: It is capable of advising the human being for the query of any domain from the particular ES.

Providing decision-making capabilities: It provides the capability of decision making in any domain, such as making any financial decision, decisions in medical science, etc.

Demonstrating a device: It is capable of demonstrating any new products such as its features, specifications, how to use that product, etc.

Problem-solving: It has problem-solving capabilities.

Explaining a problem: It is also capable of providing a detailed description of an input problem.

Interpreting the input: It is capable of interpreting the input given by the user.

Predicting results: It can be used for the prediction of a result.

Diagnosis: An ES designed for the medical field is capable of diagnosing a disease without using multiple components as it already contains various inbuilt medical tools.

5. Advantages of Expert System

These systems are highly reproducible.

They can be used for risky places where the human presence is not safe.

Error possibilities are less if the KB contains correct knowledge.

The performance of these systems remains steady as it is not affected by emotions, tension, or fatigue.

They provide a very high speed to respond to a particular query.

6. Limitations of Expert System

The response of the expert system may get wrong if the knowledge base contains the wrong information.

Unlike a human being, it cannot produce a creative output for different scenarios.

Its maintenance and development costs are very high.

Knowledge acquisition for designing is much difficult.

For each domain, we require a specific ES, which is one of the biggest limitations.

It cannot learn from itself and hence requires manual updates.

New Words

expert	['eksp3:t]	n.专家，能手；权威；行家
extract	['ekstrækt]	vt.提取
query	['kwiəri]	v.查询，询问
approach	[ə'prəutʃ]	n.方法；途径
suggestion	[sə'dʒestʃən]	n.建议，意见，暗示；联想，启发
performance	[pə'fɔ:məns]	n.性能；执行
understandable	[ˌʌndə'stændəbl]	adj.能懂的，可理解的
reliable	[ri'laiəbl]	adj.可靠的；可信赖的；真实可信的
format	['fɔ:mæt]	n.格式
		vt.格式化
display	[dis'plei]	v.显示

		n.显示器
deduce	[di'dju:s]	vt.推论，推断；演绎
deterministic	[di͵tɜ:mi'nistik]	adj.确定性的
probabilistic	[͵prɒbəbi'listik]	adj.概率的
uncertainty	[ʌn'sɜ:tnti]	n.无把握、不确定的事物
precise	[pri'sais]	adj.清晰的；精确的
collection	[kə'lekʃn]	n.收集，采集
formalize	['fɔ:məlaiz]	vt.使正式，形式化
symptom	['simptəm]	n.症状；征兆
participant	[pɑ:'tisipənt]	n.参加者，参与者
gather	['gæðə]	vt.收集，搜集，采集
codify	['kəʊdifai]	v.编制，整理
formalism	['fɔ:məlizəm]	n.形式主义
advice	[əd'vais]	n.建议；报告
memorize	['meməraiz]	vt.记住，记忆；熟记；存储
update	[͵ʌp'deit]	vt.更新
emotion	[i'məʊʃn]	n.情感，感情；情绪
fatigue	[fə'ti:g]	n.疲劳，疲乏
depression	[di'preʃn]	n.萎靡不振，沮丧
anxiety	[æŋ'zaiəti]	n.焦虑，忧虑
resolve	[ri'zɒlv]	vi.解决
advise	[əd'vaiz]	vt.报告；提议，建议
		vi.劝告，商量；建议，提供意见
demonstrate	['demənstreit]	vt.证明，证实；论证；显示，演示，说明
diagnosis	[͵daiəg'nəʊsis]	n.诊断
disease	[di'zi:z]	n.疾病；弊端
inbuilt	['inbilt]	adj.嵌入的，内藏的
reproducible	[͵ri:prə'dju:səbl]	adj.可再生的，能繁殖的
risky	['riski]	adj.冒险的，危险的

Phrases

inference rule	推理规则，推断规则，推理法则

be designed for...	为……而设计
search box	搜索框
period of time	一段时间
user interface	用户界面
inference engine	推理引擎，推理机
deterministic inference engine	确定性推理机，确定性推理引擎
probabilistic inference engine	概率推理机，概率推理引擎
forward chaining	正向推理；正向链接；前向链接；前向推理
backward chaining	反向推理；反向链接；后向链接；逆向推理
factual knowledge	事实性知识
bacterial infection	细菌感染
knowledge engineer	知识工程师

Abbreviations

ES (Expert System)	专家系统

Text A 参考译文
专家系统

专家系统是一种计算机程序，旨在像人类专家那样解决复杂问题并提供决策能力。它根据用户查询、使用推理和推理规则从知识库中提取知识来执行此操作。

专家系统是人工智能的一部分，第一个专家系统开发于 1970 年，这是第一个成功的人工智能方法。它通过提取存储在知识库中的知识，像人类专家那样使用事实和启发式方法帮助对复杂问题做出决策。之所以这样称呼它，是因为它包含特定领域的专家知识，并且可以解决该特定领域的任何复杂问题。这些系统是为特定领域（例如医学、科学等领域）设计的。

专家系统的性能取决于存储在其知识库中的专家的知识，知识库中存储的知识越多，系统的性能就越高。专家系统的一个常见例子是在 Google 搜索框中键入时提示拼写错误。

1. 专家系统的特点

高性能：专家系统可高效、高准确度地解决特定领域的任何类型复杂问题。

易于理解：它以用户易于理解的方式做出响应。它可以接受人类语言的输入，并以相同的方式提供输出。

可靠：非常可靠地提供高效且准确的输出。

响应速度快：专家系统可在很短的时间内为任何复杂的查询提供结果。

2. 专家系统的组成

2.1 用户界面

专家系统借助用户界面与用户进行交互，以可读格式输入查询，并将其传递给推理引擎。从推理引擎获得响应后，将结果显示给用户。换句话说，它是帮助非专家用户与专家系统进行通信以找到解决方案的界面。

2.2 推理引擎（引擎规则）

推理引擎被称为专家系统的大脑，因为它是系统的主要处理单元。它将推理规则应用于知识库，以得出结论或推论新信息，有助于得出针对用户询问的无错误解决方案。

在推理引擎的帮助下，系统从知识库中提取知识。

有两种类型的推理引擎。

确定性推理引擎：从这种类型的推理引擎得出的结论被认定是正确的，它基于事实和规则。

概率推理引擎：这种类型的推理引擎包含结论中的不确定性以及基于概率的不确定性。

推理引擎使用以下模式导出解决方案。

正向推理：从已知事实和规则开始，然后应用推理规则将其结论添加到已知事实中。

反向推理：这是一种从目标开始并向后运行以证明已知事实的向后推理方法。

2.3 知识库

知识库是一种存储类型，用于存储从特定领域的不同专家那里获得的知识，是一种大型的知识存储。知识库越大，专家系统就越精确。

知识库类似于包含特定领域或主题的信息和规则的数据库，人们还可以将知识库视为对象及其属性的集合。例如，狮子是物体，它的属性是哺乳动物，不是家畜等。

以下是知识库的组成部分。

事实知识：基于事实并被知识工程师接受的知识属于事实知识。

启发式知识：此知识基于实践，具有猜测、评估和经验的特点。

知识表述：它用于使用 if-else 规则来形式化存储在知识库中的知识。

知识获取：这是提取、组织和构造领域知识、指定从各个专家那里获取知识的规则并将该知识存储到知识库中的过程。

2.4 专家系统的开发

在这里，我们将以 MYCIN ES 为例来说明专家系统的开发过程。以下是构建 MYCIN 的一些步骤。

首先，专家系统应该具有专业知识。就 MYCIN 而言，专门从事细菌感染医学领域的人类专家提供该领域一些疾病的原因、症状和其他知识的信息。

MYCIN 的知识库已成功更新，医生提出了一个新的问题对它进行测试，问题是通过输入患者的详细信息（包括症状、当前状况和病史）来识别细菌的存在。

专家系统将需要由患者填写问卷，以了解有关患者的一般信息，例如性别、年龄等。

现在，系统已收集了所有信息，因此它将通过推理引擎和存储在知识库中的事实并应用 if-then 规则来找到问题的解决方案。

最后，它将通过用户界面为患者提供响应。

2.5 专家系统开发的参与者

专家系统的构建有三个主要参与者。

专家：专家系统的成功很大程度上取决于人类专家提供的知识，这些专家是专门研究某特定领域的人员。

知识工程师：知识工程师是从领域专家那里收集知识，然后根据一定的形式将该知识编码到系统中的人。

最终用户：是特定的某个人或一群人，他们可能不是专家，而是在专家系统上工作，需要获得查询的解决方案或建议，查询的内容很复杂。

3. 为什么要使用专家系统

在使用任何技术之前，我们必须了解为什么要使用该技术，对专家系统也一样。以下是我们为什么需要专家系统的要点。

没有内存限制：专家系统可以存储所需数量的数据，并可以在应用时提取。而对于人类专家而言，要每次都记住所有数据是很困难的。

高效：如果使用正确的知识更新知识库，专家系统将提供高效的输出，这对于人类而言也许是不可能的。

具备某个领域的专业知识：每个领域都有很多人类专家，他们都有不同的技能和不同的经验，因此要获得查询的最终输出并不容易。但是，如果我们将从人类专家那里获得的知识放到专家系统中，它将通过综合所有事实和知识来提供有效的输出。

不受情感影响：这些系统不受诸如疲劳、愤怒、沮丧、焦虑等人类情感的影响，因此，性能保持恒定。

高安全性：这些系统提供高安全性以响应任何查询。

考虑所有事实：要响应任何查询，它将检查并考虑所有可用事实并相应地提供结果，而人类专家有可能由于某些原因而没有考虑到某些事实。

定期更新可以提高性能：如果专家系统提供的结果存在问题，我们可以通过更新知识库来提高系统的性能。

4. 专家系统的能力

专家系统的能力包括：

提供建议：从特定专家系统人能够查询到任何领域的相关建议。

提供决策能力：它提供任何领域的决策能力，例如做出任何财务决策、医学决策等。

演示设备：它能够演示任何新产品，例如其功能、规格、如何使用该产品等。

解决问题的能力：它具有解决问题的能力。

说明问题：它也能够提供所输入问题的详细描述。

解释输入：能够解释用户给出的输入。

预测结果：可用于结果预测。

诊断：为医学领域设计的专家系统无须使用多个组件即可诊断疾病，因为它已经包含各种内置医疗工具。

5. 专家系统的优势

这些系统是高度可复制的。

可用于威胁人身安全的危险场所。

如果知识库包含正确的知识，则出错的可能性会较小。

由于不受情绪、紧张或疲劳的影响，这些系统的性能保持稳定。

它们对特定查询提供了很高的响应速度。

6. 专家系统的局限性

如果知识库包含错误的信息，专家系统的响应可能会出错。

它无法像人类那样针对不同场景产生创造性的输出。

它的维护和开发成本很高。

获取用于设计专家系统的知识非常困难。

对于每个领域，我们都需要特定的专家系统，这是其最大的限制之一。

它无法自学，因此需要手动更新。

Text B
Fuzzy Logic in Artificial Intelligence and Its Applications

1. What Is Fuzzy Logic

扫码听课文

Fuzzy Logic (FL) is a method of reasoning that resembles human reasoning. This approach is similar to how humans perform decision making. And it involves all intermediate possibilities between YES and NO.

The conventional logic block that a computer understands takes precise input and produces a definite output as TRUE or FALSE, which is equivalent to a human being's YES or NO. The fuzzy logic was invented by Lotfi Zadeh, who observed that unlike computers, humans have a different range of possibilities between YES and NO, such as: Certainly Yes, Possibly Yes, Cannot Say, Possibly No, Certainly No.

The fuzzy logic works on the levels of possibilities of input to achieve a definite output. It can be implemented in systems with different sizes and capabilities such as micro-controllers, large networked or workstation-based systems. It can also be implemented in hardware, software or a combination of both.

2. Why Do We Use Fuzzy Logic

Generally, we use the fuzzy logic system for both commercial and practical purposes such as:
- It controls machines and consumer products.
- If not accurate reasoning, it at least provides acceptable reasoning.
- This helps in dealing with the uncertainty in engineering.

3. Fuzzy Logic Architecture

Now that you know about fuzzy logic in AI and why we actually use it, let's move on to understand the architecture of this logic.

The architecture of fuzzy logic consists of four main parts (see Figure 5-1):
- Rules: It contains all the rules and the if-then conditions offered by the experts to control the decision-making system. The recent update in the fuzzy theory provides different effective methods for the design and tuning of fuzzy controllers. Usually, these developments reduce the number of fuzzy rules.

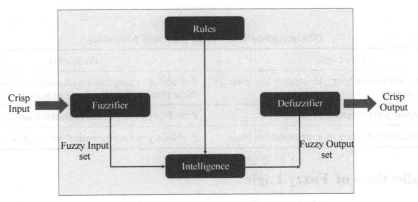

Figure 5-1 The architecture of fuzzy logic

- Fuzzification: This step converts inputs or the crisp numbers into fuzzy sets. You can measure the crisp inputs by sensors and pass them into the control system for further processing.
- Inference Engine: It determines the degree of match between fuzzy input and the rules. According to the input field, it will decide the rules that are to be fired. Combining the fired rules, form the control actions.
- Defuzzification: The defuzzification process converts the fuzzy sets into a crisp value. There are different types of techniques available, and you need to select the best-suited one with an expert system.

4. Membership Function

The membership function is a graph that defines how each point in the input space is mapped to membership value between 0 and 1. It allows you to quantify linguistic terms and represent a fuzzy set graphically. A membership function for a fuzzy set A on the universe of discourse X is defined as $\mu A: X \to [0,1]$

It quantifies the degree of membership of the element in X to the fuzzy set A.
- x-axis represents the universe of discourse.
- y-axis represents the degrees of membership in the [0, 1] interval.

There can be multiple membership functions applicable to fuzzify a numerical value. Simple membership functions are used as the complex functions do not add precision in the output.

5. Fuzzy Logic vs. Probability

There are differences between fuzzy logic and probability in AI. Some of the differences between fuzzy logic and probability are shown in Table 5-1.

Table 5-1　　　　　　　　　Differences between fuzzy logic and probability

Fuzzy Logic	Probability
In fuzzy logic, we basically try to capture the essential concept of vagueness	Probability is associated with events and not facts, and those events will either occur or not occur
Fuzzy logic captures the meaning of partial truth	Probability theory captures partial knowledge
Fuzzy logic takes truth degrees as a mathematical basis	Probability is a mathematical model of possibility

6. Applications of Fuzzy Logic

Now let's have a look at some of the applications of this logic. Fuzzy logic is used in various fields such as automotive systems, domestic goods, environment control, etc. Some of the common applications are:

- It is used in the aerospace field for altitude control of spacecraft and satellite.
- It controls the speed and traffic in the automotive systems.
- It is used for decision making support systems and personal evaluation in the large company business.
- It also controls the PH, drying, chemical distillation processes in the chemical industry.
- Fuzzy logic is used in natural language processing and various intensive applications in artificial intelligence.
- It is extensively used in modern control systems such as expert systems.
- Fuzzy logic mimics how a person would make decisions, only much faster. Thus, you can use it with neural networks.

7. Advantages & Disadvantages of Fuzzy Logic

Fuzzy logic provides simple reasoning similar to human reasoning. There are more such advantages of using this logic, such as:

- The structure of fuzzy logic systems is easy and understandable.
- Fuzzy logic is widely used for commercial and practical purposes.
- It helps you to control machines and consumer products.
- It helps you to deal with the uncertainty in engineering.
- It is mostly robust as no precise inputs are required.
- If the feedback sensor stops working, you can program it into the situation.
- You can easily modify it to improve or alter system performance.

- Inexpensive sensors can be used, which helps you to keep the overall system cost and complexity low.

Though fuzzy logic has many advantages, it has some disadvantages as well:

- Fuzzy logic is not always accurate. So the results are perceived based on assumptions and may not be widely accepted.
- It cannot recognize machine learning and neural network type patterns.
- Validation of a fuzzy knowledge-based system needs extensive testing with hardware.
- It is a difficult task to set exact fuzzy rules and membership functions.
- At times, the fuzzy logic is confused with probability theory.

New Words

fuzzy	['fʌzi]	adj.模糊的；含糊不清的
resemble	[ri'zembl]	vt.与……相像，类似于
involve	[in'vɒlv]	vt.包含；使参与
convention	[kən'venʃn]	n.惯例，习俗，规矩
definite	['definit]	adj.明确的；一定的
micro-controller	['maikrəʊ kən'trəʊlə]	n.微控制器
workstation	['wɜːksteiʃn]	n.工作站
acceptable	[ək'septəbl]	adj.可接受的；令人满意的
tune	[tjuːn]	vt.调整
controller	[kən'trəʊlə]	n.控制器
fuzzification	[fʌzifi'keiʃn]	n.模糊性
crisp	[krisp]	adj.清晰的
degree	[di'griː]	n.度数；程度
defuzzification	[ˌdiːfʌzifi'keiʃn]	n.去模糊化，清晰化
map	[mæp]	vt.映射
quantify	['kwɒntifai]	vt.确定……的数量
linguistic	[liŋ'gwistik]	adj.语言的
fuzzify	['fʌzifai]	v.模糊化
vagueness	[veignis]	n.含糊
occur	[ə'kɜː]	vi.发生；出现
partial	['pɑːʃl]	adj.部分的

ignorance	['ɪgnərəns]	n.无知
automotive	[ˌɔːtə'məʊtɪv]	adj.自动的；汽车的
aerospace	['eərəʊspeɪs]	n.航空航天
spacecraft	['speɪskrɑːft]	n.宇宙飞船，航天器
distillation	[ˌdɪstɪ'leɪʃn]	n.蒸馏（过程）；蒸馏物
intensive	[ɪn'tensɪv]	adj.加强的，强烈的
extensively	[ɪk'stensɪvli]	adv.广大地，广泛地
robust	[rəʊ'bʌst]	adj.健壮的，结实的，稳固的
modify	['mɒdɪfaɪ]	v.修改
alter	['ɔːltə]	v.改变，更改
inexpensive	[ˌɪnɪk'spensɪv]	adj.不贵的，便宜的
validation	[ˌvælɪ'deɪʃn]	n.确认；有效；校验

Phrases

be equivalent to	等于
a range of	一系列；一些；一套
workstation-based system	基于工作站的系统
a combination of ...	……的组合
consist of	由……组成；包括
fuzzy set	模糊集合
fired rule	触发规则
membership function	隶属函数
be associated with ...	与……联系
altitude control	高度控制
decision making support system	决策支持系统
be confused with ...	与……混淆

Abbreviations

FL (Fuzzy Logic)	模糊逻辑

Text B 参考译文
人工智能中的模糊逻辑及其应用

1. 什么是模糊逻辑

模糊逻辑（FL）是一种类似于人类推理的推理方法。这种方法类似于人类执行决策的方式，它涉及是和否之间的所有中间可能性。

计算机可以理解的常规逻辑块需要精确的输入，并产生确定的输出，如 TRUE 或 FALSE，相当于人类的 YES 或 NO。模糊逻辑是由洛特菲·扎德（Lotfi Zadeh）发明的，他观察到，与计算机不同，人类在"是"和"否"之间有不同的可能性范围，例如肯定是、可能是、不能判定、可能不是、确定不是。

模糊逻辑根据输入的不同级别实现确定的输出。它可以在具有不同大小和功能的系统中实现，例如微控制器、大型网络或基于工作站的系统。它也可以用于硬件、软件或两者的结合。

2. 为什么要使用模糊逻辑

通常，我们将模糊逻辑系统用于商业和实用目的，例如：
- 它控制机器和消费产品。
- 即使推理不准确，至少可以提供可接受的推理。
- 这有助于处理工程中的不确定性。

3. 模糊逻辑体系

既然你已经了解了 AI 中的模糊逻辑以及我们使用它的真正原因，那么让我们继续了解该逻辑的体系结构。

模糊逻辑的体系结构由四个主要部分组成（见图 5-1）。

（图略）

- 规则：包含专家提供的用于控制决策系统的所有规则和 if-then 条件。模糊理论的最新进展为模糊控制器的设计和调整提供了不同的有效方法，通常，这些发展减少了模糊规则的数量。
- 模糊化：此步骤将输入或清晰数字转换为模糊集。你可以通过传感器测量清晰的输入，并将其传递到控制系统中以进行进一步处理。
- 推理引擎：它确定模糊输入和规则之间的匹配程度。根据输入字段，它将决定要触发的规则，结合触发规则，形成控制动作。
- 去模糊化：去模糊化过程将模糊集转换为清晰的值。有不同类型的技术可用，你需要选择带有专家系统的最适合的技术。

4. 隶属函数

隶属函数是一个图形，它定义了输入空间中的每个点如何映射到介于 0 和 1 之间的隶属值。它允许你量化语言术语并以图形方式表示模糊集。话语 X 上的模糊集 A 的隶属函数被定义为 $\mu A: X \rightarrow [0,1]$

它将 X 中元素的隶属度量化为模糊集 A。

- x 轴表示话语范围。
- y 轴表示[0，1]间隔中的隶属度。

可以有多个适用于模糊化数值的隶属函数。人们使用简单的隶属函数，因为复杂的函数不会在输出中增加精度。

5. 模糊逻辑与概率

AI 中的模糊逻辑和概率之间存在差异。表 5-1 显示了模糊逻辑和概率之间的一些差异。

表 5-1　　　　　　　　　　　　模糊逻辑和概率之间的差异

模糊逻辑	概率论
在模糊逻辑中，我们基本上试图捕获模糊的基本概念	概率与事件有关而与事实无关，并且那些事件可能发生也可能不发生
模糊逻辑捕获了部分事实的含义	概率论捕捉部分知识
模糊逻辑将真实程度作为数学基础	概率论是可能性的数学模型

6. 模糊逻辑的应用

现在让我们看一下这种逻辑的一些应用。模糊逻辑被用于各个领域，例如自动系统、国内商品、环境控制等。一些常见的应用是：

- 在航天领域用于航天器和卫星的高度控制。
- 它控制自动系统中的速度和流量。
- 用于大公司业务中的决策支持系统和个人评估。
- 它还控制化学工业中的 pH 值、干燥、化学蒸馏过程。
- 模糊逻辑用于自然语言处理以及人工智能中的各种密集应用。
- 它广泛用于现代控制系统，例如专家系统。
- 模糊逻辑能更快地模仿人如何做出决策。因此，可以将其与神经网络一起使用。

7. 模糊逻辑的优缺点

模糊逻辑提供类似于人类推理的简单推理，使用此逻辑还有更多的优点，例如：
- 模糊逻辑系统的结构简单易懂。
- 模糊逻辑广泛用于商业和实践目的。
- 它可以帮助控制机器和消费产品。
- 帮助你应对工程中的不确定性。
- 由于不需要精确的输入，因此它非常稳固。
- 如果反馈传感器停止工作，则可以根据情况进行编程。
- 你可以轻松地对其进行修改，以改善或更改系统性能。
- 可以因此使用廉价的传感器，以帮助降低总体系统成本和复杂性。

尽管模糊逻辑有很多优点，但也有一些缺点：
- 模糊逻辑并不总是准确的，因此，结果是基于假设的，可能不会被广泛接受。
- 它无法识别机器学习和神经网络类型模式。
- 基于知识的模糊系统的验证需要使用硬件进行大量测试。
- 设置精确的模糊规则和隶属函数是一项艰巨的任务。
- 有时，模糊逻辑与概率论相互混淆。

Exercises

[Ex. 1] Answer the following questions according to Text A.

1. What is an expert system (ES)? How does it perform?
2. What are the characteristics of expert systems?
3. What is user interface?
4. What is the inference engine known as? Why?
5. What are the two types of inference engine? What modes does inference engine use to derive the solutions?
6. What is knowledge base?
7. How many primary participants are there in the building of expert systems? What are they?
8. If we put the knowledge gained from human experts into the expert system, what will the expert system do?
9. What is an ES designed for the medical field capable of?
10. What is the last but one advantage of the expert system mentioned in the passage? And what is its first limitation?

[Ex. 2] Fill in the following blanks according to Text B.

1. Fuzzy Logic (FL) is a method of reasoning that _____. And it involves _____ between _____ and _____.

2. The fuzzy logic was invented by _____ who observed that unlike computers, humans have _____ between YES and NO, such as: Certainly Yes, Possibly Yes, Cannot Say, Possibly No, Certainly No.

3. Generally, we use the fuzzy logic system for both _____ and _____ purposes. It controls _____ and _____. This helps in dealing with _____.

4. The architecture of fuzzy logic consists of four main parts. They are _____, _____, _____ and _____.

5. The membership function is _____ that defines how each point in the input space is mapped to _____ between 0 and 1. It allows you to _____ and _____.

6. In fuzzy logic, we basically try to capture _____ of vagueness. Probability is associated with _____ and _____, and those events will _____.
the essential concept, events, not facts, either occur or not occur

7. The difference between fuzzy logic and probability is that fuzzy logic takes truth degrees as _____ while probability is _____.

8. Fuzzy logic is used in the aerospace field for altitude control of _____. It mimics how a person would _____, only much faster. Thus, you can use it with _____.

9. Fuzzy logic is mostly robust as _____. If the feedback sensor stops working, you can _____.

10. Fuzzy logic is not always _____. So the results are perceived based on _____ and may not _____. Validation of a fuzzy knowledge-based system needs _____.

[Ex. 3] Translate the following terms or phrases from English into Chinese and vice versa.

1. backward chaining   1. _____
2. deterministic inference engine   2. _____
3. inference engine   3. _____
4. knowledge engineer   4. _____
5. probabilistic inference engine   5. _____
6. *n.*方法；途径   6. _____
7. *n.*收集，采集   7. _____

8.   *vt.* 推论，推断；演绎   8. _____
9.   *vt.* 提取   9. _____
10. *n.* 性能；执行   10. _____

[Ex. 4] Translate the following passages into Chinese.

Knowledge Engineering

    Knowledge engineering is a field of artificial intelligence that tries to emulate the judgment and behavior of a human expert in a given field.

    Knowledge engineering is the technology behind the creation of expert systems to assist with issues related to their programmed field of knowledge. Expert systems involve a large and expandable knowledge base integrated with a rules engine that specifies how to apply information in the knowledge base to each particular situation. The systems may also incorporate machine learning so that they can learn from experience in the same way that humans do. Expert systems are used in various fields including healthcare, customer service, financial services, manufacturing and the law.

    Using algorithms to emulate the thought patterns of a subject matter expert, knowledge engineering tries to take on questions and issues as a human expert would. Looking at the structure of a task or decision, knowledge engineering studies how the conclusion is reached. A library of problem-solving methods and a body of collateral knowledge are used to approach the issue or question. The amount of collateral knowledge can be very large. Depending on the task and the knowledge that is drawn on, the virtual expert may assist with troubleshooting, solving issues, assisting a human or acting as a virtual agent.

    Scientists originally attempted knowledge engineering by trying to emulate real experts. Using the virtual expert was supposed to get you the same answer as you would get from a human expert. This approach was called the transfer approach.

    A surprising amount of collateral knowledge is required to enable analogous reasoning and nonlinear thought. Currently, a modeling approach is used where the same knowledge and process need not necessarily be used to reach the same conclusion for a given question or issue. Eventually, it is expected that knowledge engineering will produce a specialist that surpasses the abilities of its human counterparts.

[Ex. 5] Fill in the blanks with the words given below.

representations	goals	major	engineer	techniques
conducted	frames	verified	acquisition	identify

Knowledge Acquisition

Knowledge acquisition refers to the process of extracting, structuring, and organizing domain knowledge from domain experts into a program. A knowledge ___1___ is an expert in AI language and knowledge representation who investigates a particular problem domain, determines important concepts, and creates correct and efficient ___2___ of the objects and relations in the domain.

Capturing domain knowledge of a problem domain is the first step in building an expert system. In general, the knowledge acquisition process through a knowledge engineer can be divided into four phases:

(1) Planning: The goal is to understand the problem domain, ___3___ domain experts, analyze various knowledge acquisition techniques, and design proper procedures.

(2) Knowledge extraction: The goal is to extract knowledge from experts by applying various knowledge ___4___ techniques.

(3) Knowledge analysis: The outputs from the knowledge extraction phase, such as concepts and heuristics, are analyzed and represented in formal forms, including heuristic rules, ___5___, objects and relations, semantic networks, classification schemes, neural networks, and fuzzy logic sets. These representations are used in implementing a prototype expert system.

(4) Knowledge verification: The prototype expert system containing the formal representation of the heuristics and concepts is ___6___ by the experts. If the knowledge base is incomplete or insufficient to solve the problem, alternative knowledge acquisition techniques may be applied, and additional knowledge acquisition process may be ___7___.

Many knowledge acquisition techniques and tools have been developed with various strengths and limitations. Commonly used ___8___ include interviewing, protocol analysis, repertory grid analysis, and observation.

Interviewing is a technique used for eliciting knowledge from domain experts and design requirements. The basic form involves free-form or unstructured question–answer sessions between the domain expert and the knowledge engineer. The ___9___ problem of this approach results from the inability of domain experts to explicitly describe their reasoning process and the biases involved in human reasoning. A more effective form of interviewing is called structured interviewing, which is goal-oriented and directed by a series of clearly stated ___10___. Here, experts either fill out a set of carefully designed questionnaire cards or answer questions carefully designed based on an established domain model of the problem-solving process. This technique reduces the interpretation problem inherent in the unstructured interviewing as well as the distortion caused by domain expert subjectivity.

Unit 6
Machine Learning

Text A
Supervised vs. Unsupervised Machine Learning

1. What Is Supervised Machine Learning

In supervised machine learning, you train the machine using data which is well "labeled." It means some data is already tagged with the correct answer. It can be compared to learning which takes place in the presence of a supervisor or a teacher.

扫码听课文

A supervised learning algorithm learns from labeled training data, helps you to predict outcomes for unforeseen data. Successfully building, scaling, and deploying accurate supervised machine learning data science model takes time and needs technical expertise from a team of highly skilled data scientists. Moreover, data scientist must rebuild models to make sure the insights given remains true until its data changes.

2. What Is Unsupervised Learning

Unsupervised learning is a machine learning technique, where you do not need to supervise the model. Instead, you need to allow the model to work on its own to discover information. It mainly deals with the unlabeled data.

Unsupervised learning algorithms allow you to perform more complex processing tasks compared to supervised learning. Unsupervised learning can be more unpredictable compared with deep learning and reinforcement learning methods.

3. Why Supervised Learning

- Supervised learning allows you to collect data or produce a data output from the previous experience.
- Supervised learning helps you to optimize performance criteria using experience.
- Supervised learning helps you to solve various types of real-world computation problems.

4. Why Unsupervised Learning

Here are the prime reasons for using unsupervised learning:

- Unsupervised machine learning finds all kinds of unknown patterns in data.
- Unsupervised methods help you to find features which can be useful for categorization.
- It takes place in real time, so all the input data will be analyzed and labeled in the presence of learners.
- It is easier to get unlabeled data from a computer than labeled data, which needs manual intervention.

5. How Does Supervised Learning Work

For example, you want to train a machine to help you predict how long it will take you to drive home from your workplace. Here, you start by creating a set of labeled data. This data includes:

- Weather conditions
- Time of the day
- Holidays

All these details are your inputs. The output is the amount of time it takes you to drive back home on that specific day.

You instinctively know that if it's raining outside, then it will take you longer to drive home. But the machine needs data and statistics.

Let's see now how you can develop a supervised learning model of this example which help the user to determine the commute time. The first thing it requires you to create is a training data set. This training set will contain the total commute time and corresponding factors like weather, time, etc. Based on this training set, your machine might see there's a direct relationship between the amount of rain and time it will take you to get home.

So, it ascertains that the more it rains, the longer you will be driving to get back to your home. You might also see the connection between the time you leave work and the time you'll be on the road.

The closer you're to 6 p.m. the longer time it takes you to get home. Your machine may find some of the relationships with your labeled data.

This is the start of your data model. It begins to see how rain impacts the way people drive. It also starts to see that more people travel during a particular time of day.

6. How Does Unsupervised Learning Work

Let's take the case of a baby and her family dog.

She knows and identifies her pet dog. A few weeks later a family friend brings along a dog and tries to play with the baby.

The baby has not seen this dog earlier. But she recognizes many features (2 ears, eyes, walking on 4 legs) are like her pet dog. She identifies a new animal like a dog. This is unsupervised learning, where you are not taught but you learn from the data (in this case data about a dog.) Had this been supervised learning, the family friend would have told the baby that it's a dog.

7. Types of Supervised Machine Learning Techniques

7.1 Regression

Regression technique predicts a single output value using training data.

Example: You can use regression to predict the house price from training data. The input variables will be locality, size of a house, etc.

7.2 Classification

Classification means to group the output inside a class. If the algorithm tries to label input into two distinct classes, it is called binary classification. Selecting between more than two classes is referred to as multiclass classification.

Example: Determining whether or not someone will be a defaulter of the loan.

Strengths: Outputs always have a probabilistic interpretation, and the algorithm can be regularized to avoid overfitting.

Weaknesses: Logistic regression may underperform when there are multiple or non-linear decision boundaries. This method is not flexible, so it does not capture more complex relationships.

8. Types of Unsupervised Machine Learning Techniques

Unsupervised learning problems are further grouped into clustering and association problems.

8.1 Clustering

Clustering is an important concept when it comes to unsupervised learning. It mainly deals with finding a structure or pattern in a collection of uncategorized data. Clustering algorithms will process your data and find natural clusters (groups) if they exist in the data. You can also modify how many clusters your algorithms should identify. It allows you to adjust the granularity of these groups.

8.2 Association

Association rules allow you to establish associations among data objects inside large databases. This unsupervised technique is about discovering exciting relationships between variables in large databases. For example, people that buy a new home most likely to buy new furniture.

Other examples:

- A subgroup of cancer patients grouped by their gene expression measurements.
- Groups of shopper based on their browsing and purchasing histories.
- Movie group by the rating given by movies viewers.

9. Summary

- In supervised learning, you train the machine using data which is well "labeled."
- Unsupervised learning is a machine learning technique, where you do not need to supervise the model.
- Supervised learning allows you to collect data or produce a data output from the previous experience.
- Unsupervised machine learning helps you to find all kind of unknown patterns in data.

(1) For example, you will able to determine the time taken to reach back home based on weather condition, times of the day and holiday.

(2) For example, baby can identify other dogs based on past supervised learning.

- Regression and classification are two types of supervised machine learning techniques.
- Clustering and association are two types of unsupervised learning.
- In a supervised learning model, input and output variables will be given while with unsupervised learning model, only input data will be given.

New Words

| supervise | [ˈsuːpəvaɪz] | v.监督；管理 |
| unsupervise | [ˌʌnˈsuːpəvaɪz] | v.无监督；不管理 |

labeled	['leibld]		*adj.*标记的
supervisor	['su:pəvaizə]		*n.*管理者，监督者，指导者
unforeseen	[,ʌnfɔː'siːn]		*adj.*未预见到的，无法预料的
scale	[skeil]		*v.*改变大小，调整比例
			*n.*规模；比例
deploy	[di'plɔi]		*v.*使展开；施展
rebuild	[,riː'bild]		*vt.*重建；恢复
insight	['insait]		*n.*洞察力，洞悉；直觉
discover	[di'skʌvə]		*vt.*发现
unlabelled	[,ʌn'leibld]		*adj.*无标签的，无标记的
unpredictable	[,ʌnpri'diktəbl]		*adj.*无法预言的，不可预测的
reinforcement	[,riːin'fɔːsmənt]		*n.*增强，加强
produce	[prə'djuːs]		*v.*生产；制作，制造
computation	[,kɒmpjʊ'teiʃn]		*n.*计算，估计
categorization	[,kætəgərai'zeiʃn]		*n.*编目方法，分门别类
manual	['mænjʊəl]		*adj.*人工的，手工的
			*n.*手册；指南
condition	[kən'diʃn]		*n.*状态；环境
output	['aʊtpʊt]		*v.*输出
instinctively	[in'stiŋktivli]		*adv.*本能地
statistic	[stə'tistik]		*adj.*统计的，统计学的
commute	[kə'mjuːt]		*vi.*通勤
ascertain	[,æsə'tein]		*vt.*弄清，确定，查明
connection	[kə'nekʃn]		*n.*连接；联系，关系
recognize	['rekəgnaiz]		*vt.*认出，识别
locality	[ləʊ'kæləti]		*n.*位置
class	[klɑːs]		*n.*类
binary	['bainəri]		*adj.*二进制的
multiclass	['mʌltiklɑːs]		*n.*多类
defaulter	[di'fɔːltə]		*n.*违约者；未偿还债务者
loan	[ləʊn]		*n.*贷款，借款
interpretation	[in,tɜːpri'teiʃn]		*n.*解释，说明
regularize	['regjʊləraiz]		*n.*规则化，正则化
overfitting	[,əʊvə'fitiŋ]		*n.*过拟合

underperform	[ˌʌndəpəˈfɔ:m]	v.表现不佳，工作不如预期
flexible	[ˈfleksəbl]	adj.灵活的；柔韧的
clustering	[ˈklʌstəriŋ]	n.聚类
granularity	[ˌgrænjʊˈlærəti]	n.粒度；间隔尺寸
association	[əˌsəʊʃiˈeiʃn]	n.联合，联系
establish	[iˈstæbliʃ]	vt.建立，创建
subgroup	[ˈsʌbgru:p]	n.小群，子群
expression	[ikˈspreʃn]	n.表达，表现，表示
measurement	[ˈmeʒəmənt]	n.量度；尺寸

Phrases

machine learning	机器学习
compare ... to	把……比作
take place	发生；举行；进行
data science	数据科学
technical expertise	技术专长
make sure	确保
supervised learning	监督学习
unsupervised learning	无监督学习
unlabeled data	无标记数据，未标记数据
labeled data	已标记数据
training data set	训练数据集
data model	数据模型
uncategorized data	未分类的数据
association rule	关联规则

Text A 参考译文
有监督机器学习与无监督机器学习

1. 什么是有监督机器学习

在有监督的机器学习中，你使用"标记"完好的数据来训练机器，这意味着某些数据已经用正确答案进行了标记。可以将其比作在主管或老师在场的情况下进行的学习。

有监督学习算法可从标记了的训练数据中学习,帮助你预测不可预见的数据的结果。成功构建、扩展和部署精确的监督机器学习数据科学模型需要时间,并且需要高技能数据科学家团队的技术专长。此外,数据科学家必须重建模型,以确保所提供的见解在数据更改之前保持真实。

2. 什么是无监督学习

无监督学习是一种机器学习技术,你无需监督模型。相反,你需要允许模型自行工作以发现信息。它主要处理未标记的数据。

与有监督学习相比,无监督学习算法可使你执行更复杂的处理任务。与深度学习和强化学习方法相比,无监督学习更不可预测。

3. 为什么要进行有监督学习

- 有监督学习使你可以从以前的经验中收集数据或产生数据输出。
- 有监督学习帮助你根据经验优化性能标准。
- 有监督学习可帮助你解决各种类型的实际计算问题。

4. 为什么要进行无监督学习

以下是使用无监督学习法的主要原因:
- 无监督机器学习可发现数据中的各种未知模式。
- 无监督方法可帮助你找到可用于分类的功能。
- 它是实时发生的,因此所有输入数据都要在学习者在场的情况下进行分析和标记。
- 从计算机获取未标记的数据比获取需要手动干预标记的数据容易。

5. 有监督学习如何进行

例如,你想训练一台机器来帮助你预测从工作场所开车回家要花费多长时间,这时,你需要首先创建一组标记数据。这些数据包括:
- 天气状况
- 一天中的具体时段
- 假期

所有这些详细信息是你的输入,输出则是在特定日期开车回家的时间。

你本能地知道,如果外面下雨,那将需要更长的时间才能到家。但是机器需要数据和统计信息。

现在让我们看一下如何开发此示例的有监督学习模型,以帮助用户确定通勤时间。你首先需要创建一个训练数据集。该训练集将包含总通勤时间和相应的因素（例如天气、时间段等）。根据此训练集，机器可能会发现雨量和回家所需的时间之间存在直接关系。

因此，可以确定下雨的时间越长，开车回家的时间就越长。它还可能会发现你下班时间和在路上的时间之间的联系。

离下午六点钟越近，回家所需的时间越长。机器可能会发现某些与标签数据有关的关系。

这是你的数据模型的开始，它看到了降雨如何影响人们的驾驶方式，还有监督到了在一天的特定时间路上有更多的人。

6. 无监督学习如何运作

让我们以一个婴儿和她家的狗为例。

这个婴儿认识她家的狗并知道它是一条狗。几周后，家人的一个朋友带来了一条狗，并试着与婴儿玩耍。

这个婴儿以前没有见过这条狗。但是她发现了很多特征（两只耳朵、两只眼睛、四只脚走路）很像她家的狗。她发现了一条像狗的新动物。这就是无监督学习，没人教你，你从数据（在本例中为关于狗的数据）中学习。如果是有监督学习，则家人的朋友会告诉婴儿这是条狗。

7. 有监督机器学习技术的类型

7.1 回归

回归技术使用训练数据预测单个输出值。

示例：你可以使用回归从训练数据中预测房价，输入变量将是位置、房屋大小等。

7.2 分类

分类是将输出分组到一个类中。如果该算法尝试将输入标记为两个不同的类，则称为二进制分类。在两个以上的类别之间进行选择称为多类别分类。

示例：确定某人是否将成为该笔贷款的拖欠人。

优点：输出始终具有概率解释，并且可以对算法进行正则化以避免过拟合。

缺点：当存在多个或非线性决策边界时，逻辑回归可能表现不佳。此方法不灵活，因此无法捕获更复杂的关系。

8. 无监督机器学习技术的类型

无监督学习问题可进一步分为聚类和关联问题。

8.1 聚类

当谈及无监督学习时，聚类是一个重要的概念，它主要涉及在未分类数据的集合中查找结构或模式。聚类算法将处理你的数据并查找自然聚类（组），如果它们存在于数据中的话。你还可以修改算法应识别的集群数量，它允许你调整这些组别的粒度。

8.2 关联

关联规则允许你在大型数据库内部的数据对象之间建立关联，这种无监督的技术能够发现大型数据库中变量之间令人兴奋的关系。例如，购买新房的人最有可能购买新家具。

其他例子：
- 按基因表达式测量结果分组的癌症患者子集。
- 基于购物者的浏览和购买历史记录的群体。
- 按电影观众给出的等级对电影分组。

9. 总结

- 在"有监督学习"中，使用"标记"完好的数据训练机器。
- 无监督学习是一种机器学习技术，无需监督模型。
- 有监督学习使你可以从以前的经验中收集数据或产生数据输出。
- 无监督机器学习可帮助发现数据中的各种未知模式。

（1）例如，你将能够根据天气状况、一天中的时间段和假期确定返回家所需的时间。
（2）例如，婴儿可以根据过去的有监督学习来识别其他狗。

- 回归和分类是有监督机器学习技术的两种类型。
- 聚类和关联是无监督学习的两种类型。
- 在有监督学习模型中，将给出输入和输出变量；而在无监督学习模型中，将仅给出输入数据。

Text B
Machine Learning Applications in the Real World

As we move forward into the digital age, one of the modern innovations we've seen is the creation of machine learning. This incredible form of artificial intelligence is already being used in various industries and professions.

扫码听课文

1. Image Recognition

It is one of the most common machine learning applications. There are many situations where you can classify the object as a digital image. For digital images, the measurements describe the outputs of each pixel in the image.

In the case of a black and white image, the intensity of each pixel serves as one measurement. So if a black and white image has N*N pixels, the total number of pixels and hence measurement is N^2.

In the coloured image, each pixel is considered as providing 3 measurements of the intensities of 3 main colour components, ie RGB.

For face detection — The categories might be face versus no face present. There might be a separate category for each person in a database of several individuals.

For character recognition — We can segment a piece of writing into smaller images, each containing a single character. The categories might consist of the 26 letters of the English alphabet, the 10 digits, and some special characters.

2. Speech Recognition

Speech recognition (SR) is the translation of spoken words into text. It is also known as "automatic speech recognition" (ASR), "computer speech recognition", or "speech to text" (STT).

In speech recognition, a software application recognizes spoken words. The measurements in this machine learning application might be a set of numbers that represent the speech signal. We can segment the signal into portions that contain distinct words or phonemes. In each segment, we can represent the speech signal by the intensities or energy in different time-frequency bands.

Speech recognition includes voice user interfaces. Voice user interfaces are such as voice dialing, call routing, and domestic appliance control. It can also be used as simple data entry, preparation of structured documents, and speech to text processing.

3. Medical Diagnosis

ML provides methods, techniques, and tools that can help to solve a variety of diagnostic and prognostic problems in the medical domain. It is being used for the analysis of the importance of clinical parameters and of their combinations for prognosis, e.g. prediction of disease progression, for the extraction of medical knowledge for outcomes research, for therapy planning and support, and for overall patient management. ML is also being used for data analysis, such as detection of regularities in the data by appropriately dealing with imperfect data, interpretation of continuous data used in the Intensive Care Unit, and for intelligent alarming resulting in effective and efficient monitoring.

It is argued that the successful implementation of ML methods can help the integration of computer-based systems in the healthcare environment, providing opportunities to facilitate and enhance the work of medical experts and ultimately to improve the efficiency and quality of medical care.

4. Statistical Arbitrage

In finance, statistical arbitrage refers to automated trading strategies that are typical of a short-term and involve a large number of securities. In such strategies, the user tries to implement a trading algorithm for a set of securities on the basis of quantities such as historical correlations and general economic variables. These measurements can be cast as a classification or estimation problem. The basic assumption is that prices will move towards a historical average.

We apply machine learning methods to obtain an index arbitrage strategy. In particular, we employ linear regression and support vector regression (SVR) onto the prices of an exchange-traded fund and a stream of stocks. By using principal component analysis (PCA) to reduce the dimension of feature space, we observe the benefit and note the issues in the application of SVR. To generate trading signals, we model the residuals from the previous regression as a mean reverting process.

In the case of classification, the categories might be sold, buying or doing nothing for each security. In the case of estimation one might try to predict the expected return of each security over a future time horizon. In this case, one typically needs to use the estimates of the expected return to make a trading decision (buy, sell, etc.).

5. Learning Associations

Learning association is the process of developing insights into various associations between products. A good example is how seemingly unrelated products may reveal an association between one another.

One application of machine learning is studying the association between the products people buy, which is also known as basket analysis. If a buyer buys "X", would he or she be forced to buy "Y" because of a relationship that can identify between them? This leads to the relationship that exists between fish and chips etc. When a new product is launched in the market, it develops a new relationship. Knowing these relationships could help in suggesting the associated product to the customer for a higher likelihood of the customer buying it. It can also help in bundling products for a better sale.

6. Classification

Classification is a process of placing each individual from the population under study in many classes. This is identified as independent variables.

Classification helps analysts to use measurements of an object to identify the category to which that object belongs. Analysts use data to establish an efficient rule. Data consists of many examples of objects with their correct classification.

For example, before a bank decides to disburse a loan, it assesses customers on their ability to repay the loan. We can do it by considering factors such as customer's earning, age, savings and financial history. This information is taken from the past data of the loan. Hence, it is used to create a relationship between customer attributes and related risks.

7. Prediction

Consider the example of a bank computing the probability of any of loan applicants faulting the loan repayment. To compute the probability of the fault, the system will first need to classify the available data in certain groups. It is described by a set of rules prescribed by the analysts.

Once we do the classification, we can compute the probability as per need. These probability computations can compute across all sectors for various purposes.

Prediction is one of the hottest machine learning algorithms. Let's take an example of retail. In the past, we were able to get insights like sales report last month or last year. These type of reporting is called historical reporting. But currently business is more interested in finding out what will be my sales next month or next year, so that it can make a required decision related to procurement, stock, etc. on time.

8. Extraction

Information extraction (IE) is another application of machine learning. It is the process of extracting structured information from unstructured data. For example web pages, articles, blogs, business reports, and e-mails. The relational database maintains the output produced by the information extraction.

The process of extraction takes input as a set of documents and produces a structured data. This output is in a summarized form such as an excel sheet and table in a relational database.

9. Regression

We can apply machine learning to regression as well.

Assume that $x = x_1, x_2, x_3, \ldots x_n$ are the input variables and y is the outcome variable. In this case, we can use machine learning technology to produce the output (y) on the basis of the input variables (x). You can use a model to express the relationship between various parameters as below:

$y = g(x)$ where g is a function that depends on specific characteristics of the model.

In regression, we can use the principle of machine learning to optimize the parameters, cut the approximation error and calculate the closest possible outcome.

We can also use machine learning for function optimization. We can choose to alter the inputs to get a better model. This gives a new and improved model to work with. This is known as response surface design.

Machine learning is an incredible breakthrough in the field of artificial intelligence.

New Words

incredible	[inˈkredəbl]	adj.不可思议的
digital	[ˈdidʒitl]	adj.数字的
intensity	[inˈtensəti]	n.强度
pixel	[ˈpiksl]	n.像素
separate	[ˈsepəreit]	v.(使)分开，分离；分割；划分
character	[ˈkærəktə]	n.字母；特点
segment	[ˈsegmənt]	n.部分，段落
band	[bænd]	n.带；波段
domestic	[dəˈmestik]	adj.家用的
prognostic	[prɒgˈnɒstik]	adj.预兆的
clinical	[ˈklinikl]	adj.临床的
prognosis	[prɒgˈnəusis]	n.[医]预后
progression	[prəˈgreʃn]	n.发展，进展
regularity	[ˌregjuˈlærəti]	n.规则性，规律性
imperfect	[imˈpɜ:fikt]	adj.有缺点的，不完美的
ultimately	[ˈʌltimətli]	adv.最后，最终
arbitrage	[ˈɑ:bitrɑ:ʒ]	n. & v.套汇，套利
security	[siˈkjuərəti]	n.有价证券
stock	[stɒk]	n.股票
residual	[riˈzidjuəl]	n.残差；剩余 adj.残余的；残留的
seemingly	[ˈsi:miŋli]	adv.看来似乎；表面上看来；貌似
unrelated	[ˌʌnriˈleitid]	adj.无关的，不相关的
population	[ˌpɒpjuˈleiʃn]	n.人口
analyst	[ˈænəlist]	n.分析师
disburse	[disˈbɜ:s]	v.支出，付出

assess	[əˈses]	vt. 评定；估价；确定金额
repay	[riˈpei]	vt. 偿还；回报
report	[riˈpɔːt]	n. 报告
blog	[blɒg]	n. 博客
unstructured	[ʌnˈstrʌktʃəd]	adj. 非结构化的
conversion	[kənˈvɜːʃn]	n. 变换，转变；转换
express	[ikˈspres]	vt. 表达
characteristic	[ˌkærəktəˈristik]	adj. 特有的；独特的 n. 特性，特征，特色
approximation	[əˌprɒksiˈmeiʃn]	n. 接近；近似值；粗略估计
breakthrough	[ˈbreikθruː]	n. 突破；重要技术成就

Phrases

serve as	充当，担任
consider ... as	把……看成
time-frequency band	时频波带，时频波段
voice user interface	语音用户接口，语音用户界面
computer-based system	基于计算机的系统
a stream of	一串串的，一连串
learning association	学习关联
in relation to ...	与……有关
basket analysis	购物篮分析
conditional probability	条件概率
be identified as ...	被识别为……，被确认为……
historical reporting	历史报告
structured information	结构化信息
unstructured data	非结构化数据
relational database	关系式数据库
structured data	结构化数据
as soon as	立刻，一经
real time	实时
approximation error	近似误差
response surface design	响应面设计

Abbreviations

SR (Speech Recognition)   语音识别
ASR (Automatic Speech Recognition)   自动语音识别
STT (Speech to Text)   语音转文本
SVR (Support Vector Regression)   支持向量回归
PCA (Principal Component Analysis)   主成分分析
IE (Information Extraction)   信息提取
RDBMS (Relational Database Management System)   关系数据库管理系统

Text B 参考译文
现实世界中机器学习的应用

迈向数字时代之后，我们已经看到的现代创新之一是机器学习的创建，这种奇妙的人工智能形式已经在各行各业中广泛应用。

1. 图像识别

这是最常见的机器学习应用之一，在许多情况下，都可以将对象分类为数字图像。对于数字图像，测量值描述了图像中每个像素的输出。

如果是黑白图像，测量的是每个像素的亮度。因此，如果黑白图像具有 $N*N$ 个像素，则像素总数（因此测量值）为 N^2。

在彩色图像中，每个像素被视为提供了 3 种主要颜色成分，即 RGB（红绿蓝）的 3 种测量值。

- 用于面部检测——可能分为面部检测和无面部检测。在几个人的数据库中，每个人可能会有一个单独的类别。
- 用于字符识别——我们可以将一段文字分割成较小的图像，每个图像包含一个字符。可能由英文的 26 个字母、10 个数字和一些特殊字符组成。

2. 语音识别

语音识别（SR）是将口头话语翻译为文本的功能，也称为"自动语音识别"（ASR），"计算机语音识别"或"语音转文本"（STT）。

在语音识别中，软件应用程序识别口头话语，此机器学习应用中测量的可能是一组代表语音信号的数字。我们可以将信号分割成包含不同单词或音素的部分，在每个片段中，我们可以通过不同时频带中的强度或能量来表示语音信号。

语音识别包括语音用户界面。语音用户界面包括语音拨号、呼叫路由和家用设备控制。它也可以用作简单的数据输入、结构化文档的准备以及语音到文本的处理。

3. 医学诊断

机器学习提供的方法、技术和工具可帮助解决医学领域的多种诊断和预后问题，可用于分析临床参数及其组合对于预后的重要性。例如，疾病进展的预测、提取医学知识以进行结果研究、制订治疗计划和支持以及对患者进行全面管理。机器学习还用于数据分析，例如通过妥当处理不完善的数据来发现数据中的规律、对重症监护室中使用的连续数据进行解释，以及进行智能警报以实现有效的监控。

有人认为，机器学习方法的成功实施可以帮助在医疗保健环境中集成基于计算机的系统，从而促进和增强医学专家的工作，并最终提高了医疗保健的效率和质量。

4. 统计套利

在金融中，统计套利是指短期内典型的涉及大量证券的自动交易策略。在这样的策略中，用户尝试基于诸如历史关联和一般经济变量之类的数量来实现针对一组证券的交易算法。这些度量可以视为分类或估计问题，基本的假设是价格将接近历史平均水平。

我们应用机器学习方法来获得指数套利策略。特别是，我们在交易所买卖基金和股票流的价格上采用线性回归和支持向量回归（SVR），通过使用主成分分析（PCA）来减少特征空间的维数，我们观察到了其中的益处并注意到了 SVR 应用中的问题。为了生成交易信号，我们将先前回归的残差建模为均值恢复过程。

在分类的情况下，每种证券的类别可能会是已出售、购买或不进行操作。人们可能会尝试预测每种证券在未来时间范围内的预期收益。在这种情况下，通常需要使用预期回报的估算值来做出交易决策（购买，出售等）。

5. 学习关联

学习关联是发现产品之间的各种关联的过程，例如揭示看似无关的产品之间的关联。

机器学习的一种应用是经常研究人们购买的产品之间的关联，这也称为购物篮分析。如果买家购买了产品"X"，由于可以识别出它与产品"Y"之间的关系，他或她会被迫购买"Y"吗？这导致了鱼和薯条等之间的关系。当新产品投放到市场时，就会建立新的关系。了解这些关系有助于向客户建议相关产品，以提高客户购买该产品的可能性，也有助于产品捆绑销售，以提高销售额。

6. 分类

分类是将所研究人群中的每个成员归入多个类别的过程，这被标识为自变量。

分类可帮助分析人员使用对象的度量值来确定该对象所属的类别。分析师使用数据来建立有效的规则，数据包含许多具有正确分类的对象示例。

例如，在银行决定发放贷款之前，它会评估客户的还款能力。通过综合考虑客户的收入、年龄、储蓄和财务历史等因素，可以做到这一点。该信息取自贷款的过去数据，因此，可以用它来建立客户属性和相关风险之间的关系。

7. 预测

以银行计算任何贷款申请人拖欠还款的可能性为例。为了计算拖欠的可能性，系统首先需要将现有数据分类为某些组，它由分析师规定的一组规则来描述。

一旦完成分类，就可以根据需要计算概率，概率计算可以针对各种目的跨所有区域进行。

预测是最热门的机器学习算法之一。让我们以零售为例，之前我们已经获得了一些洞察力，例如上个月或去年的销售报告。这种类型的报告称为历史报告。但是目前，企业对算出下个月或明年的销售额更感兴趣，这样它就可以按时做出必要的与采购、库存等有关的决定。

8. 提取

信息提取（IE）是机器学习的另一种应用。它是从非结构化数据中提取结构化信息的过程，例如从网页、文章、博客、业务报告和电子邮件中提取信息。关系数据库维护由信息提取产生的输出。

提取过程将输入作为一组文档并生成结构化数据。此输出采用汇总形式，例如关系数据库中的表格和电子表格。

9. 回归

我们也可以将机器学习应用于回归。

假设 $x = x_1, x_2, x_3, \cdots x_n$ 是输入变量，y 是结果变量。在这种情况下，我们可以使用机器学习技术根据输入变量（x）产生输出（y）。你可以使用模型来表示各种参数之间的关系，如下所示：

$y = g(x)$，其中 g 是取决于模型特定特征的函数。

在回归中，我们可以使用机器学习的原理来优化参数、减少近似误差并计算最接近的可能结果。

我们还可以将机器学习用于功能优化。可以选择更改输入以获得更好的模型，这提供了一个新的改进模型，称为响应面设计。

机器学习是人工智能领域的一项令人惊叹的突破。

Exercises

[Ex. 1] Answer the following questions according to Text A.

1. How do you train the machine in supervised machine learning? What does it mean?
2. What is unsupervised learning?
3. What is the first reason for using supervised learning and unsupervised learning respectively?
4. How do you start if you want to train a machine to help you predict how long it will take you to drive home from your workplace? What does it include?
5. How does the author show us the way unsupervised learning works?
6. What does regression technique do? What can you use regression to predict?
7. What is called binary classification? What is referred to as multiclass classification?
8. What are the strengths and weaknesses of classification?
9. What is an important concept when it comes to unsupervised learning? What does it mainly deal with?
10. What is association about? What do association rules allow you to do?

[Ex. 2] Answer the following questions according to Text B.

1. What is image recognition? What do the measurements describe for digital images?
2. What is speech recognition (SR)?
3. What does speech recognition include? What are they?
4. What does statistical arbitrage refer to in finance?
5. What do we apply machine learning methods to do? What do we employ linear regression and support vector regression (SVR) onto in particular?
6. What is learning association?
7. What does classification help analysts to do?
8. What will the system do to compute the probability of the fault?
9. What is information extraction (IE)?
10. What can we use the principle of machine learning to do in regression?

[Ex. 3] Translate the following terms or phrases from English into Chinese and vice versa.

1. data model _____ 1. _____

2. machine learning
3. supervised learning
4. unsupervised learning
5. association rule
6. *adj.* 二进制的
7. *n.* 聚类
8. *n.* 表达，表现，表示
9. *n.* 规则化，正则化
10. *v.* 无监督；不管理

2. ____
3. ____
4. ____
5. ____
6. ____
7. ____
8. ____
9. ____
10. ____

[Ex. 4] Translate the following passages into Chinese.

Advantages and Disadvantages of Machine Learning

1. Advantages of Machine Learning

1.1 Automating Time-Consuming Tasks

ML-based applications have automated several tasks like low-level decision making, data entry, tele-calling, loan approval processes.

1.2 Cost Saving

Once the algorithm is developed and put into production, it can save significant cost as human labor and decision making is minimal.

1.3 Turnaround Time

For a lot of applications, total time is of paramount importance. ML has been able to reduce time in domains such as auto insurance claims where user uploads pictures and insurance amount gets calculated. It has also helped e-commerce companies in handling returns of inventory sold.

1.4 Data-Driven Decision Making

Not only companies but a lot of governments are relying on ML to make decisions in deciding which projects to invest in and how to optimally utilize existing resources.

2. Disadvantages of Machine Learning

2.1 ML algorithms Can Be Biased

Lots of times, input data to the ML algorithm is biased to a specific gender, race, country, caste, etc. This results in ML algorithms propagating unwanted bias into the decision-making process. This has been observed in some applications which deployed ML-like social media recommendations.

2.2 ML Requires Large Data to Achieve Acceptable Accuracy

While people can learn easily for small data sets, for some applications, machine learning requires huge amounts of data to achieve sufficient accuracy.

[Ex. 5] Fill in the blanks with the words given below.

| analysis | pages | form | unrelated | challenge |
| relationship | available | purposes | report | examining |

Machine Learning Applications

1. Learning Associations

Learning association is the process of developing insights into various associations between products. A good example is how seemingly __1__ products may reveal an association between one another.

Learning association is one application of machine learning—often studying the association between the products people buy, which is also known as basket __2__. If a buyer buys 'X', will he or she buy 'Y' because of a relationship between them? This leads to the relationship that exists between fish and chips etc. When new products are launched in the market, a new __3__ is developed. Knowing these relationships can help in suggesting the associated products to the customer and it can also help in bundling products for a better package.

This learning of associations between products by a machine is learning associations. Once we find an association by __4__ a large amount of sales data, big data analysts can develop a rule to derive a probability test in learning a conditional probability.

2. Prediction

Consider the example of a bank computing the probability of any of loan applicants faulting the loan repayment. To compute the probability of the fault, the system will first need to classify the __5__ data in certain groups. It is described by a set of rules prescribed by the analysts.

Once we do the classification, we can compute the probability as per need. These probability computations can compute across all sectors for various __6__.

The current prediction is one of the hottest machine learning algorithms. Let's take an example of retail. In the past we were able to get insights from sales __7__ of last month or last year. These type of reporting is called historical reporting. But currently business is more interested in finding out what will be my sales next month or year so that it can take a required decision (related to procurement, stocks, etc.) on time.

3. Extraction

Information Extraction (IE) is another application of machine learning. It is the process of extracting structured information from unstructured data, for example web ___8___ , articles, blogs, business reports, and e-mails. The relational database maintains the output produced by the information extraction.

The process of extraction takes input as a set of documents and produces a structured data. This output is in a summarized ___9___ such as an excel sheet and table in a relational database.

Nowadays extraction is becoming a key in the big data industry. At present a huge volume of data is getting generated, most of which is unstructured. The main ___10___ is handling unstructured data. Now the conversion of unstructured data to structured data is based on some patterns.

Unit 7
Artificial Neural Network

Text A
Artificial Neural Network

扫码听课文

1. What Is Artificial Neural Network (ANN)

The term "artificial neural network" is derived from biological neural networks that develop the structure of a human brain. Similar to the human brain that has neurons interconnected to one another, artificial neural networks also have neurons that are interconnected to one another in various layers of the networks. These neurons are known as nodes.

The following figure illustrates the typical diagram of biological neural network (see Figure 7-1).

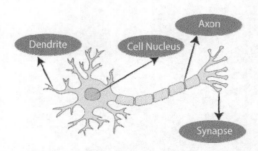

Figure 7-1 The typical diagram of biological neural network

The typical artificial neural network looks something like the following figure (see Figure 7-2).

Dendrites from biological neural network represent inputs in artificial neural networks, cell nucleus represents nodes, synapse represents weights, and axon represents output (see Table 7-1).

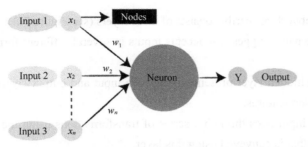

Figure 7-2　The typical artificial neural network

Table 7-1　Relationship between biological neural network and artificial neural network

Biological Neural Network	Artificial Neural Network
dendrites	inputs
cell nucleus	nodes
synapse	weights
axon	output

An artificial neural network attempts to mimic neurons of a human brain so that computers will understand things and make decisions in a human-like manner. The artificial neural network is designed by programming computers to behave simply like interconnected brain cells.

There are around 1,000 billion neurons in the human brain. Each neuron has an association point somewhere in the range of 1,000 and 100,000. In the human brain, data is stored in such a manner as to be distributed, and we can extract more than one piece of this data when necessary from our memory in parallel. We can say that the human brain is made up of incredibly amazing parallel processors.

We can understand the artificial neural network with an example. Consider an example of a digital logic gate that takes an input and gives an output. "OR" gate takes two inputs. If one or both the inputs are "On," then we get "On" in output. If both the inputs are "Off", then we get "Off" in output. Here the output depends upon input. Our brain does not perform the same task. The outputs to inputs relationship keeps changing because of the neurons in our brain which are "learning."

2. The Architecture of Artificial Neural Network

To understand the concept of the architecture of artificial neural network, we have to understand what a neural network consists of. In order to define a neural network that consists of a large number of artificial neurons which are termed units arranged in a sequence of layers, let us look at various types of layers available in an artificial neural network.

Artificial neural network primarily consists of three layers (see Figure 7-3):

Input layer: As the name suggests, it accepts inputs in several different formats provided by the programmer.

Hidden layer: The hidden layer presents in-between input and output layers. It performs all the calculations to find hidden features.

Output layer: The input goes through a series of transformations using the hidden layer, which finally results in output that is conveyed using this layer.

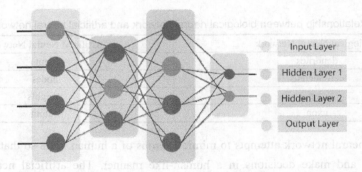

Figure 7-3　Three layers of artificial neural network

The artificial neural network takes input and computes the weighted sum of the inputs and includes a bias. This computation is represented in the form of a transfer function.

$$\sum_{i=1}^{n} Wi*Xi + b$$

It determines weighted total being passed as an input to an activation function to produce the output. Activation functions choose whether a node should fire or not. Only those who are fired make it to the output layer. There are distinctive activation functions available that can be applied to the sort of task we are performing.

3. Advantages of Artificial Neural Network

3.1　Parallel Processing Capability

Artificial neural networks have a numerical value that can perform more than one task simultaneously.

3.2　Storing Data on the Entire Network

Data that is used in traditional programming is stored on the whole network, not on a database. The disappearance of a couple of pieces of data in one place doesn't prevent the network from working.

3.3 Capability to Work with Incomplete Knowledge

After ANN training, the information may produce output even with inadequate data. The loss of performance here relies upon the significance of missing data.

3.4 Having Fault Tolerance

Failure of one or more cells of ANN does not prohibit it from generating output, and this feature makes the network fault-tolerance.

4. Disadvantages of Artificial Neural Network

4.1 Difficulty in Determining Network Structure

There is no particular guideline for determining the structure of artificial neural networks. The appropriate network structure is accomplished through experience, trial, and error.

4.2 Unrecognized Behavior of the Network

It is the most significant issue of ANN. When ANN produces a testing solution, it does not provide insight concerning why and how. It decreases trust in the network.

4.3 Hardware Dependence

Artificial neural networks need processors with parallel processing power, as per their structure. Therefore, the realization of the equipment is dependent.

4.4 Difficulty of Showing the Issue to the Network

ANNs can work with numerical data. Problems must be converted into numerical values before being introduced to ANN. The presentation mechanism to be resolved here will directly impact the performance of the network. It relies on the user's abilities.

5. How Do Artificial Neural Networks Work

Artificial neural network can be best represented as a weighted directed graph (see Figure 7-4), where the artificial neurons form the nodes. The association between the neurons outputs and neuron inputs can be viewed as the directed edges with weights. The artificial neural network receives the input signal from the external source in the form of a pattern and image in the form of a vector. These inputs are then mathematically assigned by the notations $x(N)$ for every n number of inputs.

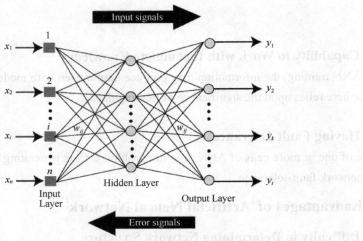

Figure 7-4 How artificial neural networks work

Then, each of the input is multiplied by its corresponding weights (these weights are the details utilized by the artificial neural networks to solve a specific problem). In general terms, these weights normally represent the strength of the interconnection between neurons inside the artificial neural network. All the weighted inputs are summarized inside the computing unit.

The activation function refers to the set of transfer functions used to achieve the desired output. There are different kinds of activation functions, but primarily either linear or non-linear sets of functions. Some of the commonly used sets of activation functions are the sigmoid, tanh and ReLU.

6. Types of Artificial Neural Network

There are various types of artificial neural networks depending upon the human brain neuron and network functions. The majority of the artificial neural networks will have some similarities with a more complex biological partner and are very effective at their expected tasks. For example, segmentation or classification.

6.1 Feedback ANN

In this type of ANN, the output returns into the network to accomplish the best-evolved results internally. The feedback networks feed information back into itself and are well suited to solve optimization issues. The internal system error corrections utilize feedback ANNs.

6.2 Feedforward ANN

Feedforward neural network is a type of artificial neural network. It adopts a unidirectional multilayer structure, where each layer contains several neurons. In this type of neural network, each neuron can receive the signal of the neuron in the previous layer and generate the output to the next layer. The 0th layer is called the input layer, the last layer is called the output layer, and the other intermediate layers are called hidden layers. The hidden layer can be one layer or multiple layers.

There is no feedback in the entire network, and the signal transmits unidirectionally from the input layer to the output layer, which can be represented by a directed acyclic graph.

The feedforward neural network has a simple structure and is widely used. It can approximate any continuous function and square integrable function with arbitrary precision. And it can accurately realize any limited training sample set. From a system perspective, the feedforward network is a static nonlinear mapping. Through complex mapping of simple nonlinear processing units, complex nonlinear processing capabilities can be obtained. From a computational point of view, there is a lack of rich dynamic behavior. Most feedforward networks are learning networks, and their classification and pattern recognition capabilities are generally better than feedback networks.

New Words

biological	[ˌbaiəˈlɒdʒikl]	adj.生物的
neural	[ˈnjʊərəl]	adj.神经的
network	[ˈnetwɜ:k]	n.网络
dendrite	[ˈdendrait]	n.树突
synapse	[ˈsainæps]	n.（神经元的）突触
weight	[weit]	n.权重
axon	[ˈæksɒn]	n.（神经的）轴突
cell	[sel]	n.细胞
parallel	[ˈpærəlel]	adj.平行的，并行的
distribute	[diˈstribju:t]	vt.分布；分配；散布
processor	[ˈprəʊsesə]	n.处理器
gate	[geit]	n.门
architecture	[ˈɑ:kitektʃə]	n.体系结构
define	[diˈfain]	v.定义
arrange	[əˈreindʒ]	v.安排；排列
accept	[əkˈsept]	vt.接受
programmer	[ˈprəʊɡræmə]	n.程序员，程序设计者

calculation	[ˌkælkjʊˈleɪʃn]	n.计算
transformation	[ˌtrænsfəˈmeɪʃn]	n.转换，变换
convey	[kənˈveɪ]	vt.传达，传递
bias	[ˈbaɪəs]	n.偏差
distinctive	[dɪˈstɪŋktɪv]	adj.有特色的，与众不同的
sort	[sɔːt]	n.& v.分类；排序
simultaneously	[ˌsɪmlˈteɪniəsli]	adv.同时地
disappearance	[ˌdɪsəˈpɪərəns]	n.消失，不见；失踪
inadequate	[ɪnˈædɪkwət]	adj.不充足的
significance	[sɪgˈnɪfɪkəns]	n.意义，意思；重要性
guideline	[ˈgaɪdlaɪn]	n.指导方针；指导原则
appropriate	[əˈprəʊpriət]	adj.适当的，恰当的，合适的
accomplish	[əˈkʌmplɪʃ]	vt.完成；达到
unrecognize	[ʌnˈrekəgnaɪz]	vt.无法识别，无法辨认
significant	[sɪgˈnɪfɪkənt]	adj.重要的；显著的
decrease	[dɪˈkriːs]	v.减少，降低
dependence	[dɪˈpendəns]	n.依赖，依靠
realization	[ˌriːəlaɪˈzeɪʃn]	n.实现
dependent	[dɪˈpendənt]	adj.依靠的，依赖的
ability	[əˈbɪləti]	n.能力，才能
vector	[ˈvektə]	n.矢量，向量
notation	[nəʊˈteɪʃn]	n.记号，标记法
interconnection	[ˌɪntəkəˈnekʃn]	n.互相连络
positive	[ˈpɒzətɪv]	adj.正的
infinity	[ɪnˈfɪnəti]	n.无穷大
benchmark	[ˈbentʃmɑːk]	n.基准，参照 vt.检测（用基准问题测试）
linear	[ˈlɪniə]	adj.线性的
segmentation	[ˌsegmenˈteɪʃn]	n.分段；分节
classification	[ˌklæsɪfɪˈkeɪʃn]	n.分类，分级
feedback	[ˈfiːdbæk]	n.反馈
feed	[fiːd]	v.馈送，提供
optimization	[ˌɒptəmaɪˈzeɪʃn]	n.最佳化，最优化
feedforward	[ˌfiːdˈfɔːwəd]	n.前馈

adopt	[əˈdɒpt]	v.采用（某方法）
multilayer	[ˈmʌltiˈleiə]	n.多层
approximate	[əˈprɒksimət]	vi.接近于，近似于
integrable	[ˈintigrəbl]	adj.可积（分）的
arbitrary	[ˈɑ:bitrəri]	adj.任意的，随意的
sample	[ˈsɑ:mpl]	n.样本
		vt.取样
static	[ˈstætik]	adj.静态的
mapping	[ˈmæpiŋ]	v.映射
dynamic	[daiˈnæmik]	adj.动态的；动力的

Phrases

be derived from	起源于，来自
biological neural network	生物神经网络
cell nucleus	细胞核
be made up of ...	由……构成，由……组成
logic gate	逻辑门
depend upon	依赖，依靠；取决于
a sequence of	一系列的
a series of	一系列；一连串
transfer function	传递函数，转移函数
activation function	激励函数；激活函数
be applied upon	适用于
parallel processing	并行处理
numerical value	数值
a couple of	几个
fault tolerance	容错
be converted into	被转换成
weighted directed graph	加权有向图
be viewed as	被视为
be multiplied by	乘以
scale up	按比例提高，按比例增加

maximum value	最大值
unidirectional multilayer structure	单向多层结构
directed acyclic graph	有向无环图
continuous function	连续函数
square integrable function	平方可积函数
sample set	样本集合

Text A 参考译文
人工神经网络

1. 什么是人工神经网络

术语"人工神经网络"源自构成人脑结构的生物神经网络。人工神经网络类似于具有相互连接的神经元的人脑，也具有在网络的各个层中相互连接的神经元，这些神经元称为节点。

下面的图表明了生物神经网络的典型模样（见图 7-1）。

（图略）

典型的人工神经网络看起来类似以下图形（见图 7-2）。

（图略）

来自生物神经网络的树突代表人工神经网络中的输入，细胞核代表节点，突触代表权重，轴突代表输出（见表 7-1）。

表 7-1　　　　　　　　　　生物神经网络与人工神经网络的关系

生物神经网络	人工神经网络
树突	输入
细胞核	节点
同步	权重
轴突	输出

人工神经网络试图模仿人脑的神经元，这样计算机就会以类似人类的方式理解事物和做出决定。人工神经网络是通过对计算机进行编程而设计的，其行为就像互连的脑细胞一样。

人脑中大约有 1 万亿个神经元，每个神经元的关联点在 1000 到 10 万个之间。在人脑中，数据以分布的方式存储，并且在必要时可以并行地从记忆中提取多个数据。可以说，人类的大脑由令人难以置信的惊人的并行处理器组成。

我们可以通过一个例子来理解人工神经网络。例如，一个采用输入并给出输出的数字逻辑门的例子："或"门需要两个输入，如果一个或两个输入均为"On"，则输出为"On"。如果两个输入均为"Off"，则输出为"Off"。这里的输出取决于输入。我们的大脑不会执行相同的任务，由于我们大脑中的神经元正在"学习"，因此输出和输入的关系在不断发生变化。

2. 人工神经网络的架构

要了解人工神经网络架构的概念，我们必须了解神经网络的组成。为了定义由大量人工神经元——被称为按层序列排列的单元——组成的神经网络，让我们看一下人工神经网络中可用的各种类型的层。

人工神经网络主要由三层组成（见图7-3）。

输入层：顾名思义，它接受程序员提供的几种不同格式的输入。

隐藏层：隐藏层出现在输入和输出层之间，它执行所有计算以查找隐藏的特征和模式。

输出层：输入使用隐藏层进行一系列转换，最终使用该层传递输出结果。

（图略）

人工神经网络获取输入并计算输入的加权和，包含一个偏差。该计算以传递函数的形式表示。

（公式略）

它确定加权总和作为输入传递给激活函数以产生输出。激活函数选择是否触发节点，只有被触发的信息能够到达输出层。有许多独特的激活函数可以应用于我们正在执行的任务。

3. 人工神经网络的优势

3.1 并行处理能力

人工神经网络具有一个可以同时执行多个任务的数值。

3.2 在整个网络上存储数据

传统编程中使用的数据存储在整个网络中，而不是存储在数据库中。在一个地方丢失了几条数据并不会妨碍网络的正常工作。

3.3 具备使用不完整知识的工作能力

人工神经网络经过训练后，即使数据不足，也可能产生输出信息。这里的性能损失取决于丢失数据的重要性。

3.4 具有容错能力

人工神经网络的一个或多个单元的失效不会阻止其生成输出，并且此功能使网络具有容错能力。

4. 人工神经网络的缺点

4.1 难以确定网络结构

没有特别的指导方法能够确定人造神经网络的结构，适当的网络结构是通过经验、尝试和错误来实现的。

4.2 无法识别的网络行为

这是人工神经网络最重要的问题。当 ANN 生成测试解决方案时，它不会提供有关原因和方式的见解，这降低了网络的可信度。

4.3 硬件依赖性

人工神经网络的结构，使得它需要具有并行处理能力的处理器。因此，它依赖相关设备来实现。

4.4 难以向网络表明问题

人工神经网络可以处理数字数据。在将问题引入 ANN 之前，必须将其转换为数值。这里要解决的表示机制将直接影响网络的性能。它取决于用户的能力。

5. 人工神经网络如何工作

用加权有向图（见图 7-4）表示人工神经网络最得当，其中人工神经元形成节点，神经元输出和神经元输入之间的关联可以被视为具有权重的有向边。人工神经网络以模式形式接收来自外部源的输入信号，以矢量形式接收图像。然后，这些输入由符号 $x(N)$ 对每 n 个输入数进行数学分配。

（图略）

然后，将每个输入乘以其相应的权重（这些权重是人工神经网络用来解决特定问题的详细信息）。一般而言，这些权重通常表示人工神经网络内部神经元之间的互连强度。所有加权输入都在计算单元内部汇总。

激活函数是指用于实现所需输出的一组传递函数。激活函数的类型不同，但主要是线性或非线性函数集。一些常用的激活函数集为 sigmoid、tanh 和 ReLU。

6. 人工神经网络的类型

根据人脑神经元和网络功能把人工神经网络分为各种类型。大多数人工神经网络与更复杂的生物伙伴（人）具有某些相似之处，并且在完成预期任务方面非常有效。例如，细分或分类

6.1 反馈人工神经网络

在这种类型的人工神经网络中,输出返回给网络,以便在网络内部得到最完善的结果。反馈网络将信息反馈回自身,非常适合解决优化问题。内部系统错误校正利用反馈人工神经网络。

6.2 前馈人工神经网络

前馈神经网络是人工神经网络的一种。前馈神经网络采用一种单向多层结构,其中每一层包含若干个神经元。在此种神经网络中,每个神经元可以接收前一层神经元的信号,并产生输出到下一层。第 0 层叫作输入层,最后一层叫作输出层,其他中间层叫作隐藏层。隐藏层可以是一层,也可以是多层。

在整个网络中无反馈,信号从输入层向输出层单向传播,可用一个有向无环图表示。

前馈神经网络结构简单,应用广泛,能够以任意精度逼近任意连续函数及平方可积函数. 而且可以精确实现任意有限训练样本集。从系统的角度来看,前馈网络是一种静态非线性映射。通过简单非线性处理单元的复合映射,可获得复杂的非线性处理能力。从计算方面看,它缺乏丰富的动力学行为。大部分前馈网络都是学习网络,其分类能力和模式识别能力一般都好于反馈网络。

Text B
Deep Learning

扫码听课文

Deep learning is a type of machine learning (ML) and artificial intelligence (AI) that imitates the way humans gain certain types of knowledge. Deep learning is an important element of data science, which includes statistics and predictive modeling. It is extremely beneficial to data scientists who are tasked with collecting, analyzing and interpreting large amounts of data; deep learning makes this process faster and easier.

At its simplest, deep learning can be thought of a way to automate predictive analytics. While traditional machine learning algorithms are linear, deep learning algorithms are stacked in a hierarchy of increasing complexity and abstraction.

To understand deep learning, imagine a toddler whose first word is dog. The toddler learns what a dog is or is not by pointing to objects and saying the word dog. The parent says, "Yes, that is a dog," or, "No, that is not a dog." As the toddler continues to point to objects, he becomes more aware of the features that all dogs possess. What the toddler does, without knowing it, is to clarify a complex abstraction — the concept of dog — by building a hierarchy in which each level of abstraction is created with knowledge that was gained from the preceding layer of the hierarchy.

1. How Does Deep Learning Work

Computer programs that use deep learning go through much the same process as the toddler learning to identify the dog. Each algorithm in the hierarchy applies a nonlinear transformation to its input and uses what it learns to create a statistical model as output. Iterations continue until the output has reached an acceptable level of accuracy.

In traditional machine learning, the learning process is supervised, and the programmer has to be extremely specific when telling the computer what types of things it should be looking for to decide if an image contains a dog or does not contain a dog. This is a laborious process called feature extraction, and the computer's success rate depends entirely upon the programmer's ability to accurately define a feature set for "dog".

The advantage of deep learning is that the program builds the feature set by itself without supervision. Unsupervised learning is not only faster, but it is usually more accurate.

Initially, the computer program might be provided with training data — a set of images for which a human has labeled each image "dog" or "not dog" with meta tags. The program uses the information it receives from the training data to create a feature set for "dog" and build a predictive model. In this case, the model the computer first creates might predict that anything in an image that has four legs and a tail should be labeled "dog". Of course, the program is not aware of the labels "four legs" or "tail." It will simply look for patterns of pixels in the digital data. With each iteration, the predictive model becomes more complex and more accurate.

Unlike the toddler, who will take weeks or even months to understand the concept of "dog", a computer program that uses deep learning algorithms only needs to be shown a training set, the program will sort through millions of images, and accurately identify which images have dogs in them within a few minutes.

To achieve an acceptable level of accuracy, deep learning programs require access to immense amounts of training data and processing power, neither of which were easily available to programmers until the era of big data and cloud computing.

2. Deep Learning Methods

Various methods can be used to create strong deep learning models. These techniques include learning rate decay, transfer learning, training from scratch and dropout.

Learning rate decay. The learning rate is a hyperparameter — a factor that defines the system or sets conditions for its operation prior to the learning process — that controls how much change the model experiences in response to the estimated error every time the model weights are altered.

Learning rates that are too high may result in unstable training processes or the learning of a suboptimal set of weights. Learning rates that are too low may produce a lengthy training process that has the potential to get stuck.

The learning rate decay method — also called learning rate annealing or adaptive learning rates — is the process of adapting the learning rate to increase performance and reduce training time. The easiest and most common adaptations of learning rate during training include techniques to reduce the learning rate over time.

Transfer learning. This process involves perfecting a previously trained model; it requires an interface to the internals of a preexisting network. Firstly, users feed the existing network new data containing previously unknown classifications. Once adjustments are made to the network, new tasks can be performed with more specific categorizing abilities. This method has the advantage of requiring much less data than others, thus reducing computation time to minutes or hours.

Training from scratch. This method requires a developer to collect a large labeled data set and configure a network architecture that can learn the features and model. This technique is especially useful for new applications, as well as applications with a large number of output categories. However, overall, it is a less common approach, as it requires large amounts of data, causing training to take days or weeks.

Dropout. This method attempts to solve the problem of overfitting in networks with large amounts of parameters by randomly dropping units and their connections from the neural network during training. It has been proven that the dropout method can improve the performance of neural networks on supervised learning tasks in areas such as speech recognition, document classification and computational biology.

3. Examples of Deep Learning Applications

Because deep learning models process information in ways similar to the human brain, they can be applied to many tasks people do. Deep learning is currently used in most common image recognition tools, natural language processing and speech recognition software. These tools are starting to appear in applications as diverse as self-driving cars and language translation services.

4. What Is Deep Learning Used for

Use cases today for deep learning include all types of big data analytics applications, especially those focused on natural language processing, language translation, medical diagnosis, stock market trading signals, network security and image recognition.

Specific fields in which deep learning is currently being used include the following:

- Customer experience. Deep learning models are already being used for chatbots. And, as it continues to mature, deep learning is expected to be implemented in various businesses to improve the customer experiences and increase customer satisfaction.
- Text generation. Machines are being taught the grammar and style of a piece of text and are then using this model to automatically create a completely new text matching the proper spelling, grammar and style of the original text.
- Industrial automation. Deep learning is improving worker safety in environments like factories and warehouses by providing services that automatically detect when a worker or object is getting too close to a machine.
- Adding color. Color can be added to black and white photos and videos using deep learning models, which was a very time-consuming manual process.
- Medical research. Cancer researchers have started implementing deep learning into their practice as a way to automatically detect cancer cells.
- Computer vision. Deep learning has greatly enhanced computer vision, providing computers with extreme accuracy for object detection and image classification, restoration and segmentation.

New Words

gain	[geɪn]	v. 获得
beneficial	[ˌbeniˈfɪʃl]	adj. 有利的，有益的
clarify	[ˈklærɪfaɪ]	v. 使清楚
nonlinear	[nɒnˈlɪniə]	adj. 非线性的
iteration	[ˌɪtəˈreɪʃn]	n. 迭代，循环
accuracy	[ˈækjʊrəsi]	n. 精确（性），准确（性）
dropout	[ˈdrɒpaʊt]	n. 随机失活
hyperparameter	[ˈhaɪpəpəˌræmɪtə]	n. 超参数
unstable	[ʌnˈsteɪbl]	adj. 不稳固的；易变的
suboptimal	[sʌbˈɒptɪməl]	adj. 未达最佳标准的；不太理想的；不太令人满意的
anneal	[əˈniːl]	n. 退火；焖火 vt. 使退火
adaptive	[əˈdæptɪv]	adj. 适应的
preexist	[ˌpriːɪɡˈzɪst]	v. 先前存在

randomly	['rændəmli]	adv.随机地，随便地
extreme	[ikˈstri:m]	adj.极端的，极限的，非常的

Phrases

a type of	一种
be tasked with	承担……任务
statistical model	统计模型
feature extraction	特征提取
meta tag	元标签，元标记
predictive model	预测模型
learning rate decay	学习率衰减
in response to...	对……做出反应
get stuck	被卡住，被困住
transfer learning	迁移学习
computational biology	计算生物学
language translation service	语言翻译服务
use case	用例
customer experience	客户体验
original text	原始文本，原文

Text B 参考译文
深度学习

　　深度学习是一种机器学习（ML）和人工智能（AI），它模仿人类获得某些类型知识的方式。深度学习是数据科学的一个重要元素，其中包括统计和预测建模。对于负责收集、分析和解释大量数据的数据科学家而言，这是极为有益的。深度学习使此过程变得更快速和更容易。

　　简单来说，深度学习可以被视为自动化预测分析的一种方式。虽然传统的机器学习算法是线性的，但深度学习算法却以越来越复杂和抽象的层次结构堆叠在一起。

　　要了解深度学习，请想象一个幼儿，他学的第一个词是狗。幼儿通过指向物体并说出"狗"一词来表明他对狗的了解。父母回应说："是，那是狗"，或者："不，那不是狗"。随着幼儿继续指向物体,他逐渐意识到所有狗所具有的特征。幼儿在不知情的情况下通过建立一个层次结构来阐明复杂的抽象概念(狗的概念),在该层次结构中,每个抽象层次都使用从上一层获得的知识来创建。

1. 深度学习如何进行

使用深度学习的计算机程序与幼儿学习识别狗的过程几乎相同。层次结构中的每个算法都对其输入进行非线性变换,并使用其学来的知识来创建统计模型作为输出。一直进行迭代直到输出结果达到可接受的精度为止。

在传统的机器学习中,学习过程是受监督的,而且当决定图像包含一只狗还是不包含一只狗时,程序员必须明确地告诉计算机应该寻找的东西的类型。这是一个费时费力的过程,称为"特征提取",计算机的成功率完全取决于程序员准确定义"狗"特征集的能力。

深度学习的优势是程序无须监督即可自行构建功能集。无监督学习不仅更快,而且通常更准确。

最初,可能会向计算机程序提供训练数据——一组图像,人类已经使用元标签将图像标记为"狗"或"非狗"。该程序使用从训练数据中接收到的信息来创建"狗"的特征集并建立预测模型。在这种情况下,计算机首先创建的模型可能会预测图像中具有四条腿和一条尾巴的任何物体都应标记为"狗"。当然,该程序不知道标签"四条腿"或"尾巴",它仅仅是查找数字数据中的像素模式。每次迭代,预测模型都会变得更加复杂和准确。

与需要花费数周甚至数月才能了解"狗"的概念的幼儿不同,只需向使用深度学习算法的计算机程序显示一个训练集,该程序就会对数百万张图像进行分类,并在几分钟内就能准确识别哪些包含狗的图像。

为了达到可接受的准确度,深度学习程序需要访问大量的训练数据和处理能力,在大数据和云计算时代之前,程序员是无法轻松做到这些的。

2. 深度学习方法

可以使用各种不同的方法来创建强大的深度学习模型。这些技术包括学习率衰减、迁移学习、从头开始训练和随机失活训练。

学习率衰减。学习速率是一个超参数——一个在学习过程之前定义系统或为其操作设置条件的因素——根据每次更改的模型权重、预先估计的误差,它控制模型经历多大的变化。太高的学习率可能会导致不稳定的训练过程或学习不太理想的权重集。学习率太低可能会导致冗长的训练过程,从而有可能被卡住。

学习速率衰减方法(也称为学习速率退火或自适应学习速率)是调整学习速率以提高性能并减少训练时间的过程。训练期间最简单、最常见的学习率调整包括随着时间的推移降低学习率的技术。

迁移学习。这个过程涉及完善先前训练过的模型；它需要一个能够到达现有网络内部的接口。首先，用户向现有网络提供包含以前未知分类的新数据。一旦完成对网络的调整，就可以使用更具体的分类功能执行新任务。这种方法的优点是比其他方法需要的数据更少，从而将计算时间减少到数分钟或数小时。

从头开始训练。此方法要求开发人员收集大量的标签数据集并配置具有学习功能和模型的网络体系结构。此技术对于新应用程序以及具有大量输出类别的应用程序特别有用。但是，总的来说，这是一种不太常用的方法，因为它需要大量的数据，导致训练需要几天或几周的时间。

随机失活。这种方法试图通过在训练过程中从神经网络中随机删除单元及其连接来解决具有大量参数的网络过拟合的问题。已经证明，随机失活方法可以改善神经网络在语音识别、文档分类和计算生物学等领域的监督学习任务中的性能。

3. 深度学习应用实例

由于深度学习模型以类似于人脑的方式处理信息，因此它们可以应用于人们执行的许多任务。深度学习目前用于大多数常见的图像识别工具、自然语言处理和语音识别软件中，也开始出现在无人驾驶汽车和语言翻译服务等各种应用中。

4. 深度学习有什么用

如今，深度学习的用例包括所有类型的大数据分析应用程序，特别是那些专注于自然语言处理、语言翻译、医学诊断、股市交易信号、网络安全和图像识别的应用程序。

当前正在使用深度学习的特定领域包括：
- 客户体验。深度学习模型已经用于聊天机器人，并且，随着其不断成熟，有望在各种企业中实施，以改善客户体验并提高客户满意度。
- 文本生成。将文本的语法和文风教给机器，然后使用该模型自动创建与原始文本的文字一致、语法和文风匹配的全新文本。
- 工业自动化。深度学习自动检测工人或物体何时离机器太近，以此改善工厂和仓库等环境中的工人安全性。
- 添加颜色。可以使用深度学习模型将颜色添加到黑白照片和视频中。过去，这是一个非常耗时的手动过程。
- 医学研究。癌症研究人员已开始在其实践中实施深度学习，以自动检测癌细胞。
- 计算机视觉。深度学习极大地增强了计算机视觉，为计算机提供了精度极高的对象检测以及图像分类、恢复和分割。

Exercises

[Ex. 1] Answer the following questions according to Text A.

1. What is the term "artificial neural network" derived from?
2. What do artificial neural networks also have? What are they known as?
3. What does an artificial neural network attempt to do?
4. How many layers does artificial neural network primarily consist of? What are they?
5. What are the advantages of artificial neural networks?
6. What makes the network fault-tolerance?
7. Is there any particular guideline for determining the structure of artificial neural networks? How is it accomplished?
8. What are the disadvantages of artificial neural networks?
9. What does the activation function refer to? And what are some of the commonly used sets of activation functions?
10. What are the two types of artificial neural networks mentioned in the passage?

[Ex. 2] Answer the following questions according to Text B.

1. What is deep learning?
2. What is the difference between traditional machine learning algorithms and deep learning algorithms?
3. How does the traditional machine learning work?
4. What is the advantage of deep learning?
5. What do deep learning programs require to achieve an acceptable level of accuracy?
6. What happens when learning rates are too high or too low?
7. What advantage does transfer learning have?
8. What has been proven about the dropout method?
9. Why can deep learning models be applied to many tasks people do?
10. What are the specific fields in which deep learning is currently being used?

[Ex. 3] Translate the following terms or phrases from English into Chinese and vice versa.

1. activation function _____   1. _____
2. biological neural network _____   2. _____
3. continuous function _____   3. _____

4. fault tolerance
5. parallel processing
6. *n.* 能力，才能
7. *n.* 体系结构
8. *n.* 基准，参照
9. *vt.* 分布；分配；散布
10. *n.* 反馈

4. _____
5. _____
6. _____
7. _____
8. _____
9. _____
10. _____

[Ex. 4] Translate the following passages into Chinese.

Artificial Neural Networks

1. What Are Artificial Neural Networks (ANN)

Human brains interpret the context of real-world situations in a way that computers can't. Neural networks were first developed in the 1950s to address this issue. An artificial neural network is an attempt to simulate the network of neurons that make up a human brain so that the computer will be able to learn things and make decisions in a humanlike manner. ANNs are created by programming regular computers to behave as though they are interconnected brain cells.

2. How Do Artificial Neural Networks Work

Artificial neural networks use different layers of mathematical processing to make sense of the information it's fed. Typically, an artificial neural network has anywhere from dozens to millions of artificial neurons—called units—arranged in a series of layers. The input layer receives various forms of information from the outside world. This is the data that the network aims to process or learn about. From the input unit, the data goes through one or more hidden units. The hidden unit's job is to transform the input into something the output unit can use.

3. What Are Artificial Neural Networks Used for

There are several ways artificial neural networks can be deployed, including to classify information, predict outcomes and cluster data. As the networks process and learn from data, they can classify a given data set into a predefined class. They can be trained to predict outputs that are expected from a given input and can identify a special feature of data and then classify the data by that special feature. Google uses a 30-layered neural network to power Google Photos as well as to power its "watch next" recommendations for YouTube videos. Facebook uses artificial neural networks for its DeepFace algorithm, which can recognize specific faces with 97% accuracy. It's also an ANN that powers Skype's ability to do translations in real-time.

Computers have the ability to understand the world around them in a very human-like manner thanks to the power of artificial neural networks.

[Ex. 5] Fill in the blanks with the words given below.

| hidden | complicated | connections | called | functionality |
| loops | layer | discovery | architecture | applications |

Three Types of Neural Networks

1. Feedforward Neural Networks

Feedforward neural networks are the first type of artificial neural networks and can be considered as the most commonly used ones today. These neural networks are __1__ feedforward neural networks because the flow of information through the network is unidirectional without going through __2__. Feedforward neural networks can further be classified into single-layered networks or multilayered networks, based on the presence of intermediate hidden layers. The number of layers depends on the complexity of the function that needs to be performed. The single-layered feedforward neural network consists of only two layers of neurons and no __3__ layers in between them. Multi-layered perceptrons consist of multiple hidden layers between the input and output layers, allowing for multiple stages of information processing.

Feedforward neural networks find __4__ in areas that require supervised learning, such as computer vision. They are most commonly used in object recognition and speech recognition systems.

2. Recurrent Neural Networks

Recurrent neural networks (RNN), as the name suggests, involves the recurrence of operations in the form of loops. These are much more __5__ than feedforward networks and can perform more complex tasks than basic image recognition. For instance, recurrent neural networks are usually used in text prediction and language generation. Making sense of and generating natural language involves much more complex processing than image recognition, which recurrent neural networks can perform due to their architecture. While in feedforward neural networks, __6__ only lead from one neuron to neurons in subsequent layers without any feedback, recurrent neural networks allow for connections to lead back to neurons in the same layer allowing for a broader range of operations.

However, conventional RNNs have a few limitations. They are difficult to train and have a very short-term memory, which limits their ___7___. To overcome the memory limitation, a newer form of RNN, known as LSTM or Long Short-term Memory networks are used. LSTMs extend the memory RNNs to enable them to perform tasks involving longer-term memory.

The main application areas for RNNs include natural languages processing problems such as speech and text recognition, text prediction, and natural language generation.

3. Convolutional Neural Networks

Convolutional neural networks are almost exclusively associated with computer vision applications. That's because their ___8___ is specifically suited for performing complex visual analyses. The convolutional neural network architecture is defined by a three-dimensional arrangement of neurons, instead of the standard two-dimensional array. The first layer in such neural networks is called a convolutional ___9___. Each neuron in the convolutional layer only processes the information from a small part of the visual field. The convolutional layers are followed by rectified layer units or ReLU, which enables the CNN to handle complicated information.

CNNs are mainly used in object recognition applications like machine vision and in self-driving vehicles.

While these types of artificial neural networks are the most common in today's AI applications, there are many others that are being innovated to achieve a level of functionality that is more comparable to the human brain. Every new ___10___ about our brain's working leads to a new breakthrough in the field of AI, leading to better models of neural networks. Thus, as we continue to understand our brains better, it is only a matter of time before we can reproduce the totality of our brain's functioning in computers.

Unit 8
Pattern Recognition

Text A
Pattern Recognition

扫码听课文

Pattern Recognition (PR) is the method of identifying and distinguishing the patterns from the images that are fed as input and the output are obtained in the form of patterns. There are five different phases in pattern recognition. They are sensing, segmentation, feature extraction, classification and post processing. With the help of these phases, pattern recognition can be done on any type of input like image recognition, biometric recognition, facial recognition, colours and shapes, recuperating patterns from distorted inputs, etc.

1. What Is Pattern Recognition

Pattern recognition is the practice of distinguishing the patterns using artificial intelligence and machine learning tools with algorithms. Applications such as facial expression recognition, speech recognition, medical image recognition, etc. are a part of PR systems. Two types of techniques are very prominent in the AI and ML dictionary. These are classification and clustering without which the definition is incomplete.

The raw data which acts as input is fed to be processed and then it is converted into machine-understandable codes. Then these codes undergo the training process which includes the steps of classification and clustering.

When we refer to the classification technique we assign appropriate class labels to any pattern that we want to recognize. These are dependent on the data of abstraction which is produced while being in the training sets and using specific prior known facts about the domain. Classification is an important step in supervised learning methods.

Clustering is used to partition the data for decision making. The decisions are critical in nature sometimes. This is a procedure followed in unsupervised learning.

Both classification and clustering are a typical form of the mathematical function used to signify the measurements which have been computed such that it can quantify the noteworthy patterns present in the object.

2. Features of Pattern Recognition

- It recognizes the patterns accurately with greater precision.
- It is able to recognize and classify even the unknown and unfamiliar entities.
- The accuracy can be relied upon and the shapes and objects can be explored from different angles.
- The identification is eased by various unknown patterns in the world of science.
- It is able to detect the minute differences and recover the original pattern in case of missing data.
- It helps to study and research the unknown domains of science in medical fields, data sciences, etc.

3. How Pattern Recognition Works

The working of pattern recognition depends on the various notions of supervised and unsupervised learning approaches. This whole phase cycle reveals the working of the pattern recognition approach (see Figure 8-1).

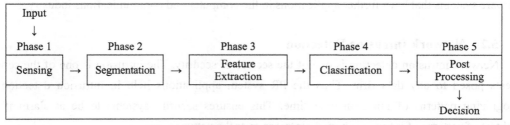

Figure 8-1 Phases of Pattern Recognition

The different phases are as follows:

Phase 1: This phase translates inputs into the analogous signal data.

Phase 2: This phase helps to isolate the sensed input data and eliminate the noise.

Phase 3: This measures the entities and the objects on its properties and sends them for further procedure of classification.

Phase 4: It assigns the sensed object to category.

Phase 5: It takes other consideration to decide for appropriate action.

4. Advantages and Disadvantages of PR Systems

Advantages:
- Pattern recognition cracks the problems in classification.
- There are various problems in day to day life that are handled by the intelligent PR systems such as facial expression recognition systems.
- Visually impaired people are also benefiting the PR systems in many domains.
- The object detection is a miraculous achievement of PR systems that is helpful in many industries such as aviation, health, etc.

Disadvantages:
- The process is quite complex and lengthy which consumes time.
- The data set needs to be large for accuracy.
- The logic is not certain of object recognition.

5. Applications of Pattern Recognition

Now let us elaborate a few applications of pattern recognition.

5.1　Natural Language Processing

The pattern recognition algorithms are used in natural language processing for building strong software systems that have further applications in the computer and communications industry.

5.2　Network Intrusion Detection

Network intrusion detection is one of the sectors of security. The intrusion is one of the serious threats posed to any data firm. Thus the PR system applications help in intrusion detection by recognizing patterns of intrusion over time. This ensures security systems to be at alarm if the slightest of patterns of intrusion show their traces over the network.

5.3　Image Sensing and Recognition

Pattern recognition well suits the image processing and its segmentation. The analysis is then performed. This is forwarded to expert reviews. PR algorithms have gradually incorporated intelligence similar to humans. Machine learning has boosted their recognition powers in medical image sensing and recognizing.

5.4 Optical Character Recognition

Optical character recognition (OCR) is a technique used in pattern recognition, artificial intelligence and machine vision. It recognizes and converts characters in images and scanned copies into editable text. With information from images and forms no longer requiring manual input, business efficiency is vastly improved and human errors reduced.

5.5 Computer Vision

Pattern recognition algorithms are widely used in computer vision. They help in extracting meaningful features from excerpts of images, videos, etc. There are applications in biomedical and medical imaging.

5.6 Disease Categorization and Healthcare

The PR systems have been employed in disease recognition and imaging over a decade.

With the arrival of sensors that can be worn on hands and easy to use devices, pattern recognition systems are performing wonderfully. The data of patients can be accessed and analyzed for the patient in almost no time and in fact in real time. PR systems are growing quickly in the healthcare industry.

5.7 Radar Signal Recognition and Analysis

Pattern recognition schemes are used in radar signal and classification. Signal processing methods are used in various applications of radar signal classifications like mine detection and identification.

5.8 Speech Recognition

The huge success of pattern recognition is seen in the speech recognition domain. The linguistics and PR systems are going hand in hand with the research and developments in various spheres of speech and linguistics recognition. It uses algorithms that are competitive and is able to treat large data sets simultaneously.

5.9 Fingerprint Identification

The fingerprint recognition technique is a dominant technology in the biometric market. A number of recognition methods have been used to perform fingerprint matching, of which pattern recognition approaches are widely used.

5.10 Agriculture

In the agriculture industry, we have a lot of applications which are reflected in the contribution of economic benefits. It works hand in hand with the breeding industry; the researchers are using multiple pattern recognition schemes for research for identification, improvement and breed key traits. This leads to dealing with the rising production demands, increasing the resistance to various diseases, and reducing the threats to the environment by using less water, fertilizers, etc.

5.11 Financial Services

In financial companies, the PR systems are helping data recognition related to trends in the financial markets. They are doing the job of identifying key insights. This might prevent financial crashes and save society from financial troubles. This technology is further used to make investments and expand businesses. Cyber surveillance is one of the examples that help in timely recognizing risks and taking steps to prevent them.

5.12 Transportation

The PR systems are expanding its applications to the transportation sector as well. On the basis of travel history, the mold of routes, packages, destinations, and costs are made pre-available to the customers using PR systems. The transportation companies thus are able to forecast potential risks that might occur on certain routes, and suitably counsel their customers timely.

New Words

distinguish	[di'stingwiʃ]	v.区分，辨别
obtain	[əb'tein]	vt.获得，得到
phase	[feiz]	n.阶段
sense	[sens]	n.感觉；意识，观念；识别力
		vt.感到；理解，领会
extraction	[ik'strækʃn]	n.取出，抽出
shape	[ʃeip]	n.形状；状态
		vt.塑造；使符合
recuperate	[ri'ku:pəreit]	vi.恢复，复原；弥补
		vt.使恢复
distort	[di'stɔ:t]	v.扭曲；变形

Unit 8 Pattern Recognition

identification	[aɪˌdentɪfɪˈkeɪʃn]	n.鉴定，识别；验明
incomplete	[ˌɪnkəmˈpliːt]	adj.不完全的；未完成的；不完备的
assign	[əˈsaɪn]	vt.分派，分配
abstraction	[æbˈstrækʃn]	n.抽象，抽象概念，抽象化
critical	[ˈkrɪtɪkl]	adj.关键的；严重的
signify	[ˈsɪgnɪfaɪ]	vt.表示……的意思；意味；预示
noteworthy	[ˈnəʊtwɜːði]	adj.值得注意的，显著的，重要的
recover	[rɪˈkʌvə]	vt.恢复；重新获得；找回
reveal	[rɪˈviːl]	vt.显露；揭露
		n.揭示，展现
analogous	[əˈnæləgəs]	adj.模拟的；相似的
consideration	[kənˌsɪdəˈreɪʃn]	n.考虑，考察；照顾，关心
impair	[ɪmˈpeə]	vt.损害，削弱
miraculous	[mɪˈrækjʊləs]	adj.奇迹般的，不可思议的
achievement	[əˈtʃiːvmənt]	n.完成，达到；成就
aviation	[ˌeɪvɪˈeɪʃn]	n.航空；飞机制造业
communication	[kəˌmjuːnɪˈkeɪʃn]	n.通讯，通信
intrusion	[ɪnˈtruːʒn]	n.入侵，闯入；干扰
detection	[dɪˈtekʃn]	n.检测，检查
threat	[θret]	n.威胁
convert	[kənˈvɜːt]	v.转变，转换
warehouse	[ˈweəhaʊs]	n.仓库；货栈；批发商店
		vt.把……放入或存入仓库
excerpt	[ˈeksɜːpt]	n.片段；摘录；节录
biomedical	[ˌbaɪəʊˈmedɪkl]	adj.生物医学的
radar	[ˈreɪdɑː]	n.雷达
sphere	[sfɪə]	n.（兴趣或活动的）范围；势力范围
		v.包围
competitive	[kəmˈpetətɪv]	adj.竞争的，比赛的
treat	[triːt]	v.对待；处理
fingerprint	[ˈfɪŋgəprɪnt]	n.指纹，指印
		vt.采指纹
dominant	[ˈdɒmɪnənt]	adj.占优势的；统治的，支配的
contribution	[ˌkɒntrɪˈbjuːʃn]	n.贡献，捐赠

trait	[treit]	n.特点，特性
resistance	[riˈzistəns]	n.抵抗；阻力，抗力；电阻
fertilizer	[ˈfɜːtəlaizə]	n.肥料，化肥
investment	[inˈvestmənt]	n.投资
cyber	[ˈsaibə]	adj.计算机（网络）的，信息技术的
transportation	[ˌtrænspɔːˈteiʃn]	n.运送，运输
mold	[məuld]	n.模子；模式；类型
		vi.对……产生影响，形成
suitably	[ˈsuːtəbli]	adv.适当地，适宜地
counsel	[ˈkaunsl]	n.协商，讨论；建议；策略
		vt.劝告，建议；提供专业咨询

Phrases

post processing	后处理
image recognition	图像识别
biometric recognition	生物识别
facial recognition	面部识别
facial expression recognition	面部表情识别
training set	训练集
decision making	决策
analogous signal	模拟信号
data set	数据集
network intrusion detection	网络入侵检测
be forwarded to	被转给
human error	人为错误
data mining	数据挖掘
knowledge discovery	知识发现
computer vision	计算机视觉
real time	实时
radar signal	雷达信号
fingerprint recognition	指纹识别
be used to	习惯于，适用于

reflect in ...	在……中反映出来
economic benefit	经济利益
hand in hand	手拉手，携手；密切合作
financial crash	金融危机
financial trouble	财务困境

Abbreviations

PR (Pattern Recognition)	模式识别
OCR (Optical Character Recognition)	光学字符识别
KDD (Knowledge Discovery in Database)	数据库中的知识发现

Text A 参考译文
模式识别

模式识别是从图像中识别和区分模式的方法，以图形作为输入，以模式的形式获得输出。模式识别有五个不同的阶段，分别是感知、分割、特征提取、分类和后处理。在这些阶段的作用下，可以在任何类型的输入上进行模式识别，例如图像识别、生物识别、面部识别、颜色和形状识别、对失真输入的复原模式等。

1. 什么是模式识别

模式识别是使用带有算法的人工智能和机器学习工具区分模式的实践活动。面部表情识别、语音识别、医学图像识别等应用是模式识别系统的一部分。在人工智能和机器学习的词典中，有两种类型的技术非常突出，这就是分类和聚类，没有它们则模式识别的定义是不完整的。

将用作输入的原始数据送入处理，然后将其转换为机器可理解的代码。之后，这些代码将经历训练过程，其中包括分类和聚类这些步骤。

当我们提到分类技术时，我们将适当的类标签分配给我们想要识别的任何模式。这依赖于抽象数据，这些数据产生于训练集中并用到该领域的特定先验事实。分类是有监督学习方法中的重要一步。

聚类用于对数据进行分区以进行决策，这些决策有时在本质上是至关重要的。这是无监督学习所遵循的过程。

分类和聚类都是数学函数的典型形式，用于表示已计算的测量值，以便可以量化对象中存在的值得注意的模式。

2. 模式识别的特征

- 它以更高的精度准确识别出模式。
- 它甚至能够对未知和陌生的实体进行识别和分类。
- 精度可靠,并且可以从不同角度探索各种形状和对象。
- 科学界的各种未知模式使鉴别变得容易。
- 它能够检测出微小的差异并在数据丢失的情况下恢复原始模式。
- 它辅助研究和探索医学、数据科学等未知科学领域。

3. 模式识别如何工作

模式识别的工作取决于有监督和无监督学习方法的各种概念,这个完整阶段周期揭示了模式识别方法的运作方式(请参见图 8-1)。

(图略)

不同的阶段如下。

阶段 1:此阶段将输入转换为模拟信号数据。

阶段 2:此阶段帮助隔离感测到的输入数据并消除噪声。

阶段 3:此阶段根据其属性测量实体和对象,并将其发送以进行进一步的分类程序。

阶段 4:此阶段将感测到的对象分类。

阶段 5:此阶段考虑其他因素以决定采取适当的措施。

4. 模式识别系统的优缺点

优点:

- 模式识别解决了分类问题。
- 诸如面部表情识别之类的智能模式识别系统可以处理日常生活中的各种问题。
- 视障人士也在许多领域中受益于模式识别系统。
- 对象检测是模式识别系统的奇迹,对航空、医疗等许多行业都有帮助。

缺点:

- 该过程非常复杂且耗时较长。
- 为了准确起见,需要海量的数据集。
- 逻辑不能确定对象识别。

5. 模式识别的应用

现在让我们详细说明模式识别的一些应用。

5.1 自然语言处理

模式识别算法用于自然语言处理中，以构建强大的软件系统，并在计算机和通信行业中进一步应用。

5.2 网络入侵检测

网络入侵检测是安全性领域之一，入侵是对任何数据公司构成的严重威胁之一。因此，模式识别系统应用程序可以通过随时识别入侵模式来帮助进行入侵检测。如果入侵模式在网络上发现蛛丝马迹，会向安全系统报警。

5.3 图像感应与识别

模式识别非常适合进行图像处理及其分割，然后执行分析，转发给专家检阅。模式识别算法已逐渐融入类似于人类的智能，机器学习增强了其在医学图像感测和识别中的识别能力。

5.4 光学字符识别

光学字符识别（OCR）是一种用于模式识别、人工智能和机器视觉的技术，它可以识别图像和扫描的副本中的字符并将其转换为可编辑的文本。由于图像和表格中的信息不再需要人工输入，因此大大提高了业务效率，减少了人为错误。

5.5 计算机视觉

模式识别算法广泛用于计算机视觉，有助于从图像、视频等的片段中提取有意义的特征，并应用在生物医学和医学成像中。

5.6 疾病分类与医疗保健

模式识别系统已经在疾病识别和医疗成像中使用了十多年。

随着可以戴在手上和易于使用的设备上的传感器的出现，模式识别系统的表现异常出色，非常快速，实际上是实时地访问和分析患者的数据。模式识别系统在医疗保健行业中发展迅速。

5.7 雷达信号识别与分析

模式识别方案用于雷达信号和分类。信号处理方法用于雷达信号分类的各种应用中，例如地雷探测和识别。

5.8 语音识别

模式识别的巨大成功体现在语音识别领域。语言学和模式识别系统与语音和语言学认识的各个领域的研究与发展齐头并进。它使用具有竞争力的算法，并且能够同时处理大型数据集。

5.9 指纹识别

指纹识别技术是生物识别市场中的主导技术,已经使用许多识别方法来执行指纹匹配,其中模式识别方法应用得很广泛。

5.10 农业

在农业领域,有许多应用,都体现在经济利益的贡献上。模式识别与育种行业携手合作;研究人员正在使用多种模式识别方案进行鉴定、改良和繁殖关键性状的研究。这有助于应对不断增长的生产需求、提高对各种疾病的抵抗力以及通过少用水和少施肥等来减少对环境的威胁。

5.11 金融服务

在金融公司中,模式识别系统正在帮助识别与金融市场趋势相关的数据。当前任务是识别关键见解,这可以防止财务崩溃,并使社会免受财务危机。该技术还用于投资和扩展业务。网络监视是帮助及时识别风险并采取措施预防风险的示例之一。

5.12 运输

模式识别系统的应用扩展到运输领域。根据旅行历史记录,使用模式识别系统的客户可以预先获得有关路线、包裹、目的地和成本的模型。因此,运输公司能够预测在某些路线上可能发生的风险,并适当地及时为其客户提供建议。

Text B
Facial Recognition

扫码听课文

Humans have always had the innate ability to recognize and distinguish between faces, yet it is only recently that computers have shown the same ability.

1. Facial Recognition Technology

Identix, a company based in Minnesota, is one of many developers of facial recognition technology. Its software, FaceIt, can pick someone's face out of a crowd, extract the face from the rest of the scene and compare it to a database of stored images. In order for this software to work, it has to know how to differentiate between a basic face and the rest of the background. Facial recognition software is based on the ability to recognize a face and then measure the various features of the face.

Every face has numerous, distinguishable landmarks, the different peaks and valleys that make

up facial features. FaceIt defines these landmarks as nodal points. Each human face has approximately 80 nodal points. Some of these measured by the software are:
- Distance between the eyes
- Width of the nose
- Depth of the eye sockets
- The shape of the cheekbones
- The length of the jaw line

These nodal points are measured to create a numerical code, called a faceprint, representing the face in the database.

In the past, facial recognition software relied on a 2D image to compare or identify another 2D image from the database. To be effective and accurate, the image captured needed to be of a face that was looking almost directly at the camera, with little variance of light or facial expression from the image in the database. This created quite a problem.

In most instances the images were not taken in a controlled environment. Even the smallest changes in light or orientation could reduce the effectiveness of the system, so they couldn't be matched to any face in the database, leading to a high rate of failure.

A newly-emerging trend in facial recognition software uses a 3D model, which claims to provide more accuracy. Capturing a real-time 3D image of a person's facial surface, 3D facial recognition uses distinctive features of the face — where rigid tissue and bone is most apparent, such as the curves of the eye socket, nose and chin — to identify the subject. These areas are all unique and don't change over time.

Using depth and an axis of measurement that is not affected by lighting, 3D facial recognition can even be used in darkness and has the ability to recognize a subject at different view angles with the potential to recognize up to 90 degrees (a face in profile).

Using the 3D software, the system goes through a series of steps to verify the identity of an individual.

1.1 Detection

Acquiring an image can be accomplished by digitally scanning an existing photograph (2D) or by using a video image to acquire a live picture of a subject (3D).

1.2 Alignment

Once it detects a face, the system determines the head's position, size and pose. As stated earlier, the subject has the potential to be recognized up to 90 degrees, while with 2D, the head must be turned at least 35 degrees toward the camera.

1.3 Measurement

The system then measures the curves of the face on a sub-millimeter scale and creates a template.

1.4 Representation

The system translates the template into a unique code. This coding gives each template a set of numbers to represent the features on a subject's face.

1.5 Matching

If the image is 3D and the database contains 3D images, then matching will take place without any changes being made to the image. However, there is a challenge currently facing databases that still contain 2D images. 3D provides a live, moving variable subject being compared to a flat, stable image. New technology is addressing this challenge. When a 3D image is taken, different points (usually three) are identified. For example, the outside of the eye, the inside of the eye and the tip of the nose will be pulled out and measured. Once those measurements are in place, an algorithm (a step-by-step procedure) will be applied to the image to convert it to a 2D image. After conversion, the software will then compare the image with the 2D images in the database to find a potential match.

1.6 Verification or Identification

In verification, an image is matched to only one image in the database (1:1). For example, an image taken of a subject may be matched to an image in the Department of Motor Vehicles database to verify the subject is who he says he is. If identification is the goal, then the image is compared to all images in the database resulting in a score for each potential match (1:N). In this instance, you may take an image and compare it to a database of mug shots to identify who the subject is.

2. Biometric Facial Recognition

The image may not always be verified or identified by facial recognition alone. Identix has created a new product to help with precision. The development of FaceIt Argus uses skinbiometrics, the uniqueness of skin texture, to yield even more accurate results.

The process, called Surface Texture Analysis, works much the same way facial recognition does. A picture is taken of a patch of skin, called a skinprint. That patch is then broken up into smaller blocks. Using algorithms to turn the patch into a mathematical, measurable space, the system will then distinguish any lines, pores and the actual skin texture. It can identify differences between identical twins, which is not yet possible using facial recognition software alone. According to Identix, by combining facial recognition with surface texture analysis, accurate identification can increase by 20 to 25 percent.

FaceIt currently uses three different templates to confirm or identify the subject: vector, local feature analysis and surface texture analysis.

- The vector template is very small and is used for rapid searching over the entire database primarily for one-to-many searching.
- The local feature analysis template performs a secondary search of ordered matches following the vector template.
- The surface texture analysis is the largest of the three. It performs a final pass after the local feature analysis template search, relying on the skin features in the image, which contains the most detailed information.

By combining all three templates, FaceIt has an advantage over other systems. It is relatively insensitive to changes in expression, including blinking, frowning or smiling and has the ability to compensate for mustache or beard growth and the appearance of eyeglasses.

However, it is not a perfect system. There are some factors that could get in the way of recognition, including:
- Significant glare on eyeglasses or wearing sunglasses
- Long hair obscuring the central part of the face
- Poor lighting that would cause the face to be over- or under-exposed
- Lack of resolution (image was taken too far away)

Identix isn't the only company with facial recognition systems available. While most work the same way FaceIt does, there are some variations. For example, a company called Animetrix, Inc. has a product called FACEngine ID SetLight that can correct lighting conditions that cannot normally be used, reducing the risk of false matches. Sensible Vision, Inc. has a product that can secure a computer using facial recognition. The computer will only power on and stay accessible as long as the correct user is in front of the screen. Once the user moves out of the line of sight, the computer is automatically secured from other users.

Due to these strides in technology, facial and skin recognition systems are more widely used than just a few years ago.

New Words

innate	[i'neit]	adj.天生的；特有的，固有的
crowd	[kraʊd]	n.人群
background	['bækgraʊnd]	n.背景；底色
numerous	['nju:mərəs]	adj.很多的，许多的
distinguishable	[di'stiŋgwiʃəbl]	adj.可区别的，可辨别的
landmark	['lændmɑ:k]	n.界标
nodal	['nəʊdəl]	adj.节的，结的
cheekbone	['tʃi:kbəʊn]	n.颊骨，颧骨
camera	['kæmərə]	n.照相机；摄影机
variance	['veəriəns]	n.变化，变动
effectiveness	[i'fektivnis]	n.有效性
curve	[kɜ:v]	n.曲线（面）
alignment	[ə'lainmənt]	n.调整
skinbiometric	[ˌskinˌbaiəʊ'metrik]	n.皮肤生物测量
uniqueness	[jʊ'ni:knis]	n.唯一性，独特性
skinprint	['skinprint]	n.皮肤纹理
insensitive	[in'sensətiv]	adj.无感觉的；不敏感的
respect	[ri'spekt]	vt.关心；遵守
glare	[gleə]	n.强光

Phrases

facial feature	面部特征
nodal point	结点，节点
eye socket	眼窝，眼眶
jaw line	下颌线
be matched to	被匹配到
pull out	拉出来
biometric facial recognition	生物面部识别
skin texture	皮肤结构，肌理
break up into	分解成

identical twin	双胞胎
surface texture analysis	表面纹理分析
local feature analysis	局部特征分析
one-to-many searching	一对多搜索，一对多检索
vector template	矢量模板

Abbreviations

2D (2-dimensional)	二维的
3D (3-dimensional)	三维的

Text B 参考译文
面部识别

人类总是天生就具有识别和区分面孔的能力，但是直到最近计算机才显示出相同的能力。

1. 面部识别技术

位于明尼苏达州的 Identix 公司是面部识别技术的众多开发商之一。它的软件 FaceIt 可以从人群中挑选出某人的脸，再从场景的其余部分中提取面部信息并将其与存储图像的数据库进行比较。为了使该软件正常工作，它必须能够区分基本面孔和背景的其余部分。面部识别软件基于识别面部然后测量面部各种特征的能力。

每张脸都有许多独特的标志和不同的峰谷构成的面部特征，FaceIt 将这些标志定义为节点，每个人脸都有大约 80 个节点。该软件测量以下数据：

- 眼睛之间的距离
- 鼻子的宽度
- 眼窝深度
- 颊骨的形状
- 下颌线的长度

对这些节点进行测量，就创建了一个数字代码，叫作面纹，表示数据库中的面部数据。

过去，面部识别软件依靠一个二维图像来比较或识别数据库中的另一个二维图像。为了做到有效和准确，捕获的图像必须是几乎直接看着相机的脸，并且要与数据库中图像的光线或面部表情几乎没有差异。这就产生了一个很大的问题。

在大多数情况下，图像不是在受控环境中拍摄的。即使光线或方向的最小变化也可能降低系统的效率，因此它们无法与数据库中的任何面孔相匹配，从而导致很高的故障率。

面部识别软件中的一种新兴趋势是使用三维模型，该模型声称可提供更高的准确性。捕捉人脸的实时三维图像时，三维人脸识别使用人脸的独特特征——最明显的是硬组织和骨头，例如眼窝、鼻子和下巴的曲线——来识别对象。这些区域都是唯一的，不会随时间而变化。

使用深度和不受光线影响的测量轴，三维面部识别甚至可以在暗环境中使用，并且能够以不同的视角识别被对象，甚至视角有可能多达90度（侧面）。

使用三维软件，系统将执行一系列步骤来验证个人身份。

1.1 检测

可以通过数字扫描现有照片（二维）或使用视频图像来获取对象的实时图片（三维）来完成图像的获取。

1.2 对齐

一旦检测到脸部，系统就会确定头部的位置、大小和姿势。如前所述，在视角为90度时都可能识别被摄对象，而在二维模式下，头部必须朝着相机至少转35度。

1.3 测量

然后，系统以亚毫米为单位测量面部曲线并创建模板。

1.4 表示

系统将模板转换为唯一的代码。该编码为每个模板提供了一组数字，以表示对象面部的特征。

1.5 匹配

如果图像是三维且数据库包含三维图像，则将进行匹配，而无须对图像进行任何更改。但是，对于那些包含二维图像的数据库，这就是个挑战。与平面、稳定的图像相比，三维提供了一个实时的、移动的可变对象。新技术正在解决这一挑战。拍摄三维图像时，会识别出不同的点（通常是三个）。例如，眼睛的外部、眼睛的内部和鼻尖将被拉出并进行测量。完成这些测量后，使用算法（逐步地）以将其转换为二维图像。转换后，软件将图像与数据库中的二维图像进行比较，以找到潜在的匹配项。

1.6 验证或识别

在验证中，一幅图像仅与数据库中的一幅图像（1∶1）匹配。例如，可以将拍摄对象的图像与机动车部门数据库中的图像进行匹配，以验证该对象就是他自称的那个人。如果目标是识别，则将图像与数据库中的所有图像进行比较，得出每个潜在匹配项的得分（1：N）。在这种情况下，你可以拍摄一张图像，并将其与数据库中的面部照片进行比较，以识别对象是谁。

2. 生物面部识别

仅使用面部识别可能无法始终对图像进行验证或识别。Identix 已经创建了一款新产品来帮助提高精确度。FaceIt Argus 的开发使用了皮肤生物测量术（皮肤纹理的独特性）来产生更准确的结果。

该过程称为"表面纹理分析"，其工作方式与面部识别非常相似。取一块皮肤图片，叫作皮纹，然后将该图片分成更小的块。使用算法将图片变成数学上可测量的空间，然后系统将区分出任何细纹、毛孔和实际皮肤纹理。它可以识别同卵双胞胎之间的差异，而仅使用面部识别软件尚无法实现。根据 Identix 的说法，通过将面部识别与表面纹理分析相结合，准确的识别率可以提高 20%～25%。

FaceIt 当前使用三种不同的模板来确认或识别对象，即矢量、局部特征分析和表面纹理分析。

- 向量模板非常小，主要用于快速搜索整个数据库，以进行一对多搜索。
- 局部特征分析模板按照矢量模板对已经排序的匹配进行二次搜索。
- 表面纹理分析是三者中最大的。局部特征分析模板搜索后，它会根据图像中包含最详细信息的皮肤特征执行最后一次遍历。

通过组合使用此三个模板，FaceIt 具有优于其他系统的优势。它对表情变化（包括眨眼、皱眉或微笑）相对不敏感，并且能够识别留了胡须以及带了眼镜的面部。

但是，这不是一个完美的系统，有一些因素可能会影响识别，包括：

- 眼镜上有强烈眩光或戴墨镜时
- 长发遮住脸部中央时
- 光线不足导致面部曝光过度或曝光不足时
- 缺乏分辨率（图像拍摄距离太远）时

Identix 并不是唯一拥有面部识别系统的公司，尽管其他产品的大多数功能都与 FaceIt 的相同，但还是有一些不同。例如，一家名为 Animetrix 的公司拥有一种名为 FACEngine ID SetLight 的产品，该产品可以纠正通常无法正常使用的照明条件，从而降低了误匹配的风险。Sensible Vision 公司的产品可以使用面部识别来保护计算机，只有正确的用户在屏幕前面时，计算机才可以开机并保持可访问状态，一旦用户移出视线，计算机就会开启自动保护，以免受其他用户的攻击。

由于这些技术进步，面部和皮肤识别系统比几年前得到了更广泛的使用。

Exercises

[Ex. 1] Answer the following questions according to Text A.
1. How many different phases are there in pattern recognition? What are they?
2. What is pattern recognition? What applications are a part of PR system's?
3. How is the identification eased?
4. What does the working of pattern recognition depend on? What does Phase 3 do?
5. What are the disadvantages of the PR system?
6. What are the pattern recognition algorithms used in natural language processing for?
7. How do the PR system applications help in intrusion detection?
8. What is optical character recognition (OCR)?
9. In what domain is the huge success of pattern recognition seen? What is a dominant technology in the biometric market?
10. What are the PR systems doing in financial companies?

[Ex. 2] Fill in the following blanks according to Text B.
1. FaceIt can _____ out of a crowd, extract the face from _____ and _____ to a database of stored images.
2. Facial recognition software is based on the ability to _____ recognize a face and then _____ measure the various features of the face.
3. FaceIt defines these landmarks as _____ . Each human face has _____ .
4. A newly-emerging trend in facial recognition software uses _____ which claims to provide _____ .
5. Using the 3D software, the system goes through a series of steps to verify the identity of an individual. They are _____, alignment, _____, _____, matching, _____ or _____ .
6. If the image is 3D and the database contains _____, then matching will take place _____ to the image. 3D provides a _____, _____ subject being compared to a flat, _____ image.
7. In verification, an image is matched to _____ in the database (1:1). If identification is the goal, then the image is compared to _____ in the database resulting in a score for _____ .
8. FaceIt currently uses three different templates to confirm or identify the subject: _____,

_____ and _____.

9. The surface texture analysis (STA) performs _____ after the LFA template search, relying on _____ in the image, which contains _____.

10. There are some factors that could get in the way of recognition, including:
- _____
- _____
- _____
- _____

[Ex. 3] Translate the following terms or phrases from English into Chinese and vice versa.

1. data mining _____   1. _____
2. data set _____   2. _____
3. facial recognition _____ 3. _____
4. fingerprint recognition _____ 4. _____
5. knowledge discovery _____ 5. _____
6. vt.分派，分配 _____   6. _____
7. v.转变，转换 _____   7. _____
8. v.区分，辨别 _____   8. _____
9. n.鉴定，识别；验明 _____ 9. _____
10. vt.获得，得到 _____   10. _____

[Ex. 4] Translate the following passages into Chinese.

Types of Robots

Mechanical bots come in all shapes and sizes to efficiently carry out the task for which they are designed. From the 0.2 millimeter-long "RoboBee" to the 200 meter-long robotic shipping vessel "Vindskip," robots are emerging to carry out tasks that humans simply can't. Generally, there are five types of robots.

1. Pre-Programmed Robots

Pre-programmed robots operate in a controlled environment where they do simple, monotonous tasks. An example of a pre-programmed robot would be a mechanical arm on an automotive assembly line. The arm serves one function — to weld a door on, or to insert a certain part into the engine — and it performs that task longer, faster and more efficiently than a human.

2. Humanoid Robots

Humanoid robots are robots that look like and/or mimic human behavior. These robots usually perform human-like activities (like running, jumping and carrying objects), and are sometimes designed to look like us, even having human faces and expressions.

3. Autonomous Robots

Autonomous robots operate independently of human operators. These robots are usually designed to carry out tasks in open environments that do not require human supervision. An example of an autonomous robot would be the Roomba vacuum cleaner, which uses sensors to roam throughout a home freely.

4. Teleoperated Robots

Teleoperated robots are mechanical bots controlled by humans. These robots usually work in extreme geographical conditions, weather, circumstances, etc. Examples of teleoperated robots are the human-controlled submarines and drones and to detect landmines on a battlefield.

5. Augmenting Robots

Augmenting robots either enhance current human capabilities or replace the capabilities a human may have lost. Some examples of augmenting robots are robotic prosthetic limbs or exoskeletons used to lift hefty weights.

[Ex. 5] Fill in the blanks with the words given below.

| drawbacks | enable | decisions | machine | maintaining |
| smart | efficiently | optimize | transportation | management |

What Artificial Intelligence and Machine Learning Can Do for A Smart City

Artificial intelligence and machine learning algorithms have increasingly become an integral part of several industries. Now they are making their way to ___1___ city initiatives, intending to automate and advance municipal activities and operations at large. Typically, a city when recognized as a smart city means that it is leveraging some kind of internet of things (IoT) and ___2___ learning machinery to glean data from various points.

A smart city has various use cases for AI-driven and IoT-enabled technology, from ___3___ a healthier environment to advancing public transport and safety. By leveraging AI and machine learning algorithms, along with IoT, a city can plan for better smart traffic solutions making sure that inhabitants get from one point to another as safely and ___4___ as possible. Machine Learning collects data from numerous points and conveys it all to a central server for further implementation and once data is collected, it has to be utilized in making a city smarter.

Machine learning generally takes the data generated by several apps such as Health MD applications, internet-enabled cars, etc. and leverages it to identify patterns and learn how to ___5___ the given set of services. Its tools are able to personalize the smart city experience by aggregating information about the most used roads in a city and then apply it to a ___6___ system.

On the other hand, machine learning and AI can be helpful in waste collection and its proper management and disposal which is a vital municipal activity in a city. Thus, the technology for smart recycling and waste ___7___ provides a sustainable waste management system. AI has the ability to understand how cities are being used and how they are functioning. It assists city planners in comprehending how the city is responding to various changes and initiatives.

In this way, AI-powered computer vision systems, for instance, could ___8___ computers to spot millions of elements of urban life in a chorus, including people, public workers, cars, accidents, fires, disasters, trash and much more. The system allows not only for autonomous monitoring but to make ___9___ based on the performance of each of these elements, changing behaviors over the course of each day or time, and responses to city systems by each element.

As AI and machine learning are transforming the way cities operate, deliver and maintain public amenities, the technologies come with some ___10___. Thus, there is a need to consider about retrofitted solutions that can hold the smart city initiatives continuing. So, the current smart city programs using AI and ML seems to advance city services and lives, including transportation, lighting, safety, connectivity, health services, among others.

Unit 9
Artificial Intelligence Software Development

Text A
Artificial Intelligence Software Development Tools

扫码听课文

1. The Importance of Artificial Intelligence Development Tools

While AI has significant potential, executing AI software development projects can be hard. You need a great deal of expertise to plan and budget such projects.

Developing an AI solution isn't a one dimensional project either, since you might need to use several ways to achieve your objectives. E.g., you might need to use machine learning (ML), natural language processing (NLP), vision, speech, and several other AI capabilities.

When you undertake a complex project like this, you need to use the right toolset, therefore, a robust set of AI development tools are important. Ideally, a robust AI development platform should offer the following capabilities:

- ML capabilities like deep learning, supervised algorithms, unsupervised algorithms, etc.;
- NLP capabilities like classification, machine translation, etc.;
- Expert systems;
- Automation;
- Vision capabilities like image recognition;
- Speech capabilities like speech-to-text and text-to-speech;

- Such a platform should also offer a robust cloud infrastructure.

2. Common Artificial Intelligence Software Development Tools

2.1 Microsoft Azure Artificial Intelligence Platform

As a cloud platform, Microsoft Azure hardly needs an introduction. Azure has made significant progress with its AI capabilities, and the Microsoft Azure AI platform is a popular choice for AI development.

The Azure AI platform offers all key AI capabilities, e.g.:
- Machine learning ;
- Vision capabilities like object recognition;
- Speech capabilities like speech recognition;
- Language capabilities like machine translation;
- Knowledge mining.

The ML capabilities of the Azure AI platform include the following:
- Azure ML, which is a Python-based automated ML service;
- Azure Databricks, which is an Apache Spark-based big data service that integrates with Azure ML;
- ONNX, which is an open source model format and runtime for ML.

Azure ML works with popular open source AI frameworks like TensorFlow.

The Azure AI platform has knowledge mining capabilities, and you can unlock insights from documents, images, and media using it. This includes the following:
- Azure search, which is a cloud search service with built-in AI;
- Form recogniser, which is an AI-powered extraction service to transform your documents and forms into usable data.

The Azure AI platform offers AI apps and agents, and you can customize them for use in your application. This includes Azure cognitive services, which offer a wide collection of domain-specific pre-trained AI models. The Azure cognitive services include AI models for the following:
- Vision
- Speech
- Language

There is a development environment for creating bots, and the Azure AI platform has templates for bots. This expedites your development.

2.2　Google Cloud Artificial Intelligence Platform

Google is yet another cloud computing giant that offers its AI platform. The Google Cloud AI platform offers all the key AI capabilities.

2.2.1　Machine Learning

With the Google Cloud AI platform, you can easily develop your ML project and deploy it to production. The Google AI platform provides an integrated tool chain for this, which expedites the development and deployment.

With this platform, you can build portable ML pipelines using Kubeflow, which is an open source platform from Google. You can deploy your ML project either on-premise or on the cloud. Cloud storage and BigQuery are the prominent options to store your data. You can access popular AI frameworks like TensorFlow.

2.2.2　Deep Learning

The Google Cloud AI platform offers pre-configured Virtual Machines (VMs) for creating deep learning applications. You can provision this VM quickly on the Google Cloud, and the deep learning VM image contains popular AI frameworks.

You can launch Google compute engine instances where TensorFlow, PyTorch, scikit-learn and other popular AI frameworks are already installed.

2.2.3　Natural language processing (NLP)

The Google Cloud AI platform has NLP capabilities, and you can use it to find out the meaning and structure of the text. You can use the Google NLP capabilities to analyze text, and the Google NLP API helps with this.

2.2.4　Speech

The Google Cloud AI platform has APIs for speech-to-text and text-to-speech capabilities.

Its speech-to-text API can help you convert audio to text, and it uses neural network models for this. The speech-to-text API supports 120 languages and their variations.

With its speech recognition capabilities, you can enable voice command-and-control features in your app, moreover, the App can transcribe audio.

On the other hand, the Google text-to-speech API enables you to create a natural-sounding speech from text. You can convert texts into audio files of popular formats like MP3 or LINEAR16.

2.2.5　Vision

Vision is another key capability of the Google Cloud AI platform, and you can use this to derive insights from your images. The Google Cloud AI platform offers its vision capabilities through REST and RPC APIs, and these APIs use pre-trained ML models.

Your App can detect objects and faces, moreover, it can read printed and handwritten texts using these APIs.

2.3 IBM Watson

IBM, the technology giant, has advanced AI capabilities, and IBM Watson is quite popular. There are already IBM Watson AI solutions specifically tailored for several industries like healthcare, oil & gas, advertising, financial services, media, Internet of Things (IoT), etc.

A key advantage of IBM Watson is that developers can use this platform to build their AI applications. It's an open AI for any cloud environment, and it's pre-integrated and pre-trained on flexible information architecture. This will expedite the development and deployment of your AI application.

IBM Watson offers the following to expedite your AI App development:
- It has developer tools like SDKs and detailed documentation for them.
- You can integrate Watson Assistant to build AI-powered conversational interfaces into your App.
- With IBM Watson, you can get Watson Discovery. It's an AI-powered search technology, and it can help your App to retrieve information that resides.
- IBM Watson has Natural Language Processing (NLP) capabilities, and it's known as Watson Natural Language Understanding (NLU).
- You can also make use of the IBM Watson Speech to Text capabilities when you build on the Watson developer platform.

IBM Watson developer resources can be useful for your AI App development team. There are SDKs for Swift, Ruby, Java, Python, Node.js, .NET, etc., therefore, you will likely find a suitable SDK for your project.

2.4 Infosys Nia

Infosys Nia is an AI platform that allows you to build AI-powered apps. It offers the following AI capabilities:
- Machine learning: Nia Advanced ML offers a broad range of ML algorithms that operate at speed and scale. It makes building high performing ML models easier.
- Contracts analysis: Nia contracts analysis capability includes ML, semantic modeling, and deep learning.
- Nia chatbot: You can build AI-powered chatbots with Nia, and your App can provide access to the enterprise knowledge repository. The App can also automate actions through a conversational interface.

- Nia data: Your AI App can integrate Nia data, a robust analytics solution.

2.5 Dialogflow

Dialogflow uses Google's infrastructure, moreover, it incorporates Google's ML capabilities. It runs on the Google cloud platform, therefore, you should be able to scale your AI App easily.

Dialogflow lets you build voice and text-based conversational interface for your App. Your App can run on web and mobile, moreover, you can connect your users on Google Assistant, Amazon Alexa, Facebook Messenger, etc.

The key capabilities offered by Dialogflow are ML, NLP, and speech.

2.6 BigML

BigML is highly focused on ML, and its development platform offers powerful ML capabilities. It provides robust ML algorithms, both for supervised and unsupervised learning.

You can implement instant access to its ML platform using its REST API, and you can do that both on-premises and on the cloud. BigML offers interpretable and exportable ML models, and this is a key advantage.

BigML offers the following features:
- It's programmable and repeatable. You can use popular languages like Python, Node.js, Ruby, Java, Swift, etc. to code your App, and BigML supports them.
- BigML helps you to automate your predictive modeling tasks.
- Deployment is flexible since you can deploy your AI App both on-premises or on the cloud.
- BigML has smart infrastructure solutions that help in scaling your App.
- BigML has robust security and privacy features.

New Words

development	[di'veləpmənt]	n.开发，发展，进化
execute	['eksikju:t]	vt.执行；履行；完成
project	['prɒdʒekt]	n.项目
budget	['bʌdʒit]	n.预算
		v.把……编入预算
dimensional	[di'menʃənəl]	adj.维的
capability	[ˌkeipə'biləti]	n.能力；容量；性能

Unit 9 Artificial Intelligence Software Development

undertake	[ˌʌndəˈteik]	vt.承担，从事；保证
platform	[ˈplætfɔ:m]	n.平台
infrastructure	[ˈinfrəstrʌktʃə]	n.基础设施；基础建设
introduction	[ˌintrəˈdʌkʃn]	n.采用，引进
mine	[main]	v.挖掘
runtime	[ˈrʌntaim]	n.运行时间
framework	[ˈfreimwɜ:k]	n.构架，框架
document	[ˈdɒkjʊmənt]	n.文档
built-in	[bilt in]	adj.嵌入的；内置的
transform	[trænsˈfɔ:m]	v.改变；变换
usable	[ˈju:zəbl]	adj.可用的；合用的；便于使用的
App	[æp]	n.计算机应用程序
agent	[ˈeidʒnt]	n.代理
customize	[ˈkʌstəmaiz]	vt.定制，客户化
pre-trained	[pri:ˈtreind]	adj.预先训练过的
environment	[inˈvairənmənt]	n.环境，外界
template	[ˈtempleit]	n.样板；模板
bot	[bɒt]	n.网上机器人；自动程序；机器人程式
expedite	[ˈekspidait]	vt.加快进展，迅速完成
giant	[ˈdʒaiənt]	n.巨人，卓越人物 adj.特大的，巨大的
portable	[ˈpɔ:təbl]	adj.轻便的；手提的
pipeline	[ˈpaiplain]	n.管道；渠道，传递途径
pre-configured	[ˌpri: kənˈfigəd]	adj.预先配置的
launch	[lɔ:ntʃ]	vt.发动；开展（活动、计划等）
variation	[ˌveəriˈeiʃn]	n.变体，变种；变化，变动
transcribe	[trænˈskraib]	vt.转录
handwrite	[ˈhændrait]	vt.用手写
tailor	[ˈteilə]	vt.调整使适应；裁剪
detailed	[ˈdi:teild]	adj.详细的，明细的
conversational	[ˌkɒnvəˈseiʃənl]	adj.会话的，对话的；双向的
interface	[ˈintəfeis]	n.界面；接口
integrate	[ˈintigreit]	v.合并，集成
semantic	[siˈmæntik]	adj.语义的

chatbot	[tʃætbɒt]	n.聊天机器人
access	[ˈækses]	vt.访问，存取
incorporate	[inˈkɔːpəreit]	vi.包含；吸收；合并；混合
web	[web]	n.万维网
mobile	[ˈməubail]	adj.可移动的
		n.手机
instant	[ˈinstənt]	n.瞬间，顷刻；此刻
		adj.立即的
interpretable	[inˈtɜːpritəbl]	adj.能说明的，能翻译的，可判断的
exportable	[eksˈpɔːtəbəl]	adj.可输出的
programmable	[ˈprəugræməbl]	adj.可编程的
repeatable	[riˈpiːtəbl]	adj.可重复的

Phrases

a great deal of	大量的；许多
machine translation	机器翻译
object recognition	目标识别；物体识别
open source	开源
cognitive service	认知服务
integrated tool	集成工具
compute engine	计算引擎
knowledge repository	知识库，知识仓库

Abbreviations

VM (Virtual Machine)	虚拟机
API (Application Programming Interface)	应用程序编程接口
IoT (Internet of Things)	物联网
SDK (Software Development Kit)	软件开发工具包
NLU (Natural Language Understanding)	自然语言理解

Text A 参考译文
人工智能软件开发工具

1. Artificial Intelligence 开发工具的重要性

尽管 AI 具有巨大的潜力，但执行 AI 软件开发项目可能会很困难。你需要大量的专业知识来为此类项目制订计划和编制预算。

开发 AI 解决方案也不是一维的项目，因为你可能需要使用多种方法来实现目标。例如，你可能需要使用机器学习（ML）、自然语言处理（NLP）、视觉、语音和其他几种 AI 功能。

当你进行这样的复杂项目时，需要使用正确的工具集，因此，一组强大的 AI 开发工具就非常重要。理想情况下，强大的 AI 开发平台应提供以下功能：

- 机器学习功能，例如深度学习、监督算法、无监督算法等。
- 自然语言处理功能，例如分类、机器翻译等。
- 专家系统。
- 自动化。
- 视觉功能，例如图像识别。
- 语音功能，例如语音转文字和文本转语音。
- 这样的平台还应该提供强大的云基础架构。

2. 常见的人工智能软件开发工具

2.1 Microsoft Azure AI 平台

作为一个云平台，Microsoft Azure 几乎不需要介绍。Azure 在其 AI 功能方面取得了长足的进步，Microsoft Azure AI Platform 是 AI 开发的流行选择。

Azure AI 平台提供了所有关键的 AI 功能，例如：

- 机器学习（ML）。
- 视觉功能，例如物体识别。
- 语音功能，例如语音识别。
- 语言功能，例如机器翻译。
- 知识挖掘。

Azure AI 平台的 ML 功能包括以下内容：

- Azure ML，这是基于 Python 的自动化 ML 服务。
- Azure Databricks，这是一个与 Azure ML 集成的基于 Apache Spark 的大数据服务。
- ONNX，这是 ML 的开源模型格式和运行时。

Azure ML 可与 TensorFlow 等流行的开源 AI 框架一起使用。

Azure AI 平台具有知识挖掘功能，你可以用它从文档、图像和媒体来发现见解。这包括以下内容：

- Azure 搜索，这是具有内置 AI 的云搜索服务。
- 表单识别器，这是一种由 AI 驱动的提取服务，用于将你的文档和表单转换为可用数据。

Azure AI 平台提供 AI 应用程序和代理，你可以定制它们以便在应用程序中使用。这包括 Azure 认知服务，该服务提供了广泛的针对特定领域的预训练的 AI 模型。Azure 认知服务包括用于以下几方面的 AI 模型：

- 视觉
- 语音
- 语言

Azure AI 平台有一个用于创建机器人的开发环境，具有用于机器人的模板。这会加快你的开发速度。

2.2 Google Cloud AI 平台

Google 是另一家提供 AI 平台的云计算巨头，Google Cloud AI 平台提供了所有关键的 AI 功能。

2.2.1 机器学习

借助 Google Cloud AI 平台，你可以轻松地开发 ML 项目并将其部署到生产环境。Google AI 平台为此提供了集成的工具链，从而加快了开发和部署。

有了这个平台，你可以使用 Kubeflow 构建可移植的 ML 管道，Kubeflow 是 Google 的开源平台。你可以在内部部署或在云上部署 ML 项目，云存储和 BigQuery 是存储数据的主要选择。你可以访问 TensorFlow 等流行的 AI 框架。

2.2.2 深度学习

Google Cloud AI 平台提供了用于创建深度学习应用程序的预配置虚拟机（VM）。你可以在 Google Cloud 上快速配置该 VM，深度学习 VM 映像包含流行的 AI 框架。

你可以启动已经安装了 TensorFlow、PyTorch、scikit-learn 和其他流行 AI 框架的 Google 计算引擎实例。

2.2.3 自然语言处理（NLP）

Google Cloud AI 平台具有 NLP 功能，你可以使用它来找出文本的含义和结构。你可以使用 Google NLP 功能来分析文本，而 Google NLP API 可以帮助你进行分析。

2.2.4 语音

Google Cloud AI 平台具有用于语音转换为文本和文本转换为语音功能的 API。

它的语音文本 API 可以帮助你将音频转换为文本，并且为此使用了神经网络模型。语音文本 API 支持 120 种语言及其变体。

借助其语音识别功能，你可以在应用程序中启用语音命令和控制功能，此外，该应用程序还可以转录音频。

另一方面，借助 Google 文本语音 API，你可以根据文本创建听起来很自然的语音。你可以将文本转换为流行格式（例如 MP3 或 LINEAR16）的音频文件。

2.2.5 视觉

视觉是 Google Cloud AI 平台的另一项关键功能，你可以使用它从图像中获取见解。Google Cloud AI 平台通过 REST 和 RPC API 提供了视觉功能，并且这些 API 使用预训练的 ML 模型。

你的 App 可以检测物体和面部，而且可以使用这些 API 读取打印的以及手写的文字。

2.3 IBM Watson

技术巨头 IBM 具有先进的 AI 功能，IBM Watson 非常受欢迎。它已经开发出专门针对医疗、石油和天然气、广告、金融服务、媒体、物联网（IoT）等多个行业量身定制的 AI 解决方案。

IBM Watson 的主要优势在于，开发人员可以使用该平台来构建其 AI 应用程序。它是适用于任何云环境的开放式 AI，并且已在灵活的信息体系结构上进行了预先集成和预先训练，这将加快 AI 应用程序的开发和部署。

IBM Watson 提供以下功能来加快你的 AI 应用程序开发：

- 它具有开发人员工具，例如 SDKs 和针对它们的详细文档。
- 你可以集成 Watson Assistant，以将基于 AI 的会话接口构建到你的应用程序中。
- 使用 IBM Watson，可以获得 Watson Discovery。这是一种基于 AI 的搜索技术，可以帮助你的 App 检索存储的信息。
- IBM Watson 具有自然语言处理（NLP）功能，被称为 Watson 自然语言理解（NLU）。
- 在 Watson 开发人员平台上构建时，还可以使用 IBM Watson 语音转文本功能。

IBM Watson 开发人员资源对 AI 应用程序开发团队很有用。有用于 Swift、Ruby、Java、Python、Node.js、.NET 等的 SDKs，因此，你很可能会找到适合你的项目的 SDKs。

2.4 Infosys Nia

Infosys Nia 是一个 AI 平台，方便你构建 AI 驱动的应用程序。它提供以下 AI 功能：

- 机器学习：Nia Advanced ML 提供了广泛的 ML 算法，这些算法可以快速、大规模地运行，使得构建高性能 ML 模型更加容易。
- 合同分析：Nia 合同分析功能包括 ML、语义建模和深度学习。
- Nia 聊天机器人：你可以使用 Nia 构建基于 AI 的聊天机器人，并且你的应用程序可以提供对企业知识库的访问。该应用程序还可以通过会话接口自动执行操作。
- Nia 数据：你的 AI 应用程序可以集成 Nia 数据，这是一个强大的分析解决方案。

2.5 Dialogflow

Dialogflow 使用 Google 的基础架构,并且整合了 Google 的 ML 功能。它运行在 Google 云平台上,因此,你应该能够轻松扩展 AI 应用程序。

Dialogflow 允许你为 App 构建基于语音和文本的会话接口。你的 App 可以在网络和移动设备上运行,此外,你还可以在 Google Assistant、Amazon Alexa、Facebook Messenger 等上面连接用户。

Dialogflow 提供的关键功能是 ML、NLP 和语音。

2.6 BigML

BigML 高度专注于 ML,其开发平台提供了强大的 ML 功能,它还提供了活跃的 ML 算法,适用于监督学习和无监督学习。

你可以使用其 REST API 实现对其 ML 平台的即时访问,并且可以在本地和云上进行访问。BigML 提供了可解释和可导出的 ML 模型,这是一个关键优势。

BigML 提供以下功能:

- 它是可编程且可重复的。你可以使用 Python、Node.js、Ruby、Java、Swift 等流行语言来编写 App 代码,BigML 支持它们。
- BigML 帮助你自动执行预测建模任务。
- 使用起来非常灵活,因为你可以在本地或云上部署 AI 应用程序。
- BigML 具有智能基础架构解决方案,可帮助你扩展应用程序。
- BigML 具有强大的安全性和隐私功能。

Text B
Programming Languages for Artificial Intelligence

Computer coding must be involved to implement any type of AI system, and there are a variety of programming languages that lend themselves to specific AI or machine learning tasks. Let's look at which programming languages will be the most beneficial for your specific use cases.

扫码听课文

Python, Java, C++, Lisp and Prolog are major AI programming languages used for artificial intelligence capable of satisfying different needs in the development and designing of different software. Each of them has its own particular strengths and weaknesses for a given project, so it is up to a developer to choose which of the languages will gratify the desired functionality and features of the application requirements.

1. Python

Python is by far the most popular programming language used in artificial intelligence today because it has easy to learn syntaxes, massive libraries and frameworks, dynamic applicability to a plethora of AI algorithms, and is relatively simple to write.

Python supports multiple orientation styles; including functional, object-oriented, and procedural. In addition, its massive community helps to keep this language at the forefront of the computer science industry.

Advantages:
- Python has a rich and extensive variety of library and tools.
- Python supports algorithm testing without having to implement them.
- Python's supporting object-oriented design increases a programmer's productivity.
- Compared to Java and C++, Python is faster in development.

Drawbacks:
- Developers accustomed to using Python face difficulty in adjusting to completely different syntax when they try using other languages for AI programming.
- Unlike C++ and Java, Python works with the help of an interpreter which makes compilation and execution slower in AI development.
- Not suitable for mobile computing.

2. C++

C++ is the fastest computer language and its speed is beneficial for AI programming projects that are time sensitive. It provides faster execution and has less response time which is applied in search engines and development of computer games. In addition, C++ allows extensive use of algorithms and is efficient in using statistical AI techniques. Another important factor is that C++ supports re-use of programs in development due to inheritance and data hiding thus efficient in time and cost saving. C++ is appropriate for machine learning and neural network.

Advantages:
- C++ is good for finding solutions for complex AI problems.
- C++ is rich in library functions and programming tools collection.
- C++ is a multi-paradigm programming that supports object-oriented principles thus useful in achieving organized data.

Drawbacks:
- Poor in multitasking; C++ is suitable only for implementing core or the base of specific

systems or algorithms.

- It follows the bottom-up approach thus, highly complex making it hard for newbies developers at using it for writing AI programs.

3. Java

Java is also a multi-paradigm language that follows object-oriented principles and the principle of Once Written Read/Run Anywhere (WORA). It is an AI programming language that can run on any platform that supports it without the need for recompilation.

Java is one of the most commonly used and not just in AI development. It derives a major part of its syntax from C and C++. Java is not only appropriate for NLP and search algorithms but also for neural networks.

Advantages:

- Java is very portable; it is easy to implement on different platforms because of Virtual Machine Technology.
- Unlike C++, Java is simple to use and even debug.
- Java has an automatic memory manager which eases the work of the developer.

Disadvantages:

- Java is, however, slower than C++, it has less speed in execution and more response time.
- Though highly portable, on older platforms, java would require dramatic changes on software and hardware to facilitate.
- Java is also a generally immature programming language for AI as there are still some developments ongoing.

4. Lisp

Lisp is another language used for artificial intelligence development. It is a family of computer programming language and is the second oldest programming language after Fortran. Lisp has developed over time to become strong and dynamic language in coding.

Some consider Lisp as the best AI programming language due to the favour of liberty it offers developers. Lisp is used in AI because of its flexibility for fast prototyping and experimentation which in turn facilitate Lisp to grow to a standard AI language. For instance, Lisp has a unique macro system which facilitates exploration and implementation of different levels of intelligence.

Lisp, unlike most AI programming languages, is more efficient in solving specific problems as it adapts to the needs of the solutions a developer is writing. It is highly suitable in inductive logic projects and machine learning.

Advantages:
- Lisp is fast and efficient in coding as it is supported by compilers instead of interpreters.
- Automatic memory manager was invented for Lisp, therefore, it has a garbage collection.
- Lisp offers specific control over systems resulting to their maximum use.

Drawbacks:
- Few developers are well acquainted with Lisp programming.
- Being an old programming language, Lisp requires configuration of new software and hardware to accommodate it use artificial intelligence.

5. Prolog

Prolog is also one of the oldest programming languages, also suitable for the development of programming AI. Like Lisp, it is also a primary computer language for artificial intelligence. It has mechanisms that facilitate flexible frameworks developers enjoy working with. It is a rule-based and declarative language as it contains facts and rules that dictate its artificial intelligence coding language.

Prolog supports basic mechanisms such as pattern matching, tree-based data structuring, and automatic backtracking essential for AI programming. Other than its extensive use in AI projects, Prolog is also used for creation of medical systems.

Advantages:
- Prolog has a built-in list handling, which is essential in representing tree-based data structures.
- Prolog is efficient for fast prototyping for AI programs.
- Prolog allows database creation simultaneous with running of the program.

Drawbacks:
- Prolog is old.
- Prolog has not been fully standardized in that some features differ in implementation, making the work of the developer cumbersome.

6. Conclusion

Like most software application development, developers are also using multiple languages to write artificial intelligence projects, but so far there is no perfect programming language that can fully match artificial intelligence projects. The choice of programming language often depends on the desired function of the artificial intelligence application.

New Words

implement	[ˈimplimənt]	vt.实施，执行；使生效，实现
		n.工具；手段
gratify	[ˈgrætifai]	vt.使高兴；使满意
massive	[ˈmæsiv]	adj.大的，大量的，大规模的
library	[ˈlaibrəri]	n.库
plethora	[ˈpleθərə]	n.过多，过剩
object-oriented	[ˈɒbdʒikt ˈɔːrientid]	adj.面向对象的
community	[kəˈmjuːnəti]	n.社区；社会团体
forefront	[ˈfɔːfrʌnt]	n.前列；第一线
extensive	[ikˈstensiv]	adj.广阔的，广大的
productivity	[ˌprɒdʌkˈtivəti]	n.生产率，生产力
interpreter	[inˈtɜːpritə]	n.解释器，解释程序
compilation	[ˌkɒmpiˈleiʃn]	n.编译
execution	[ˌeksiˈkjuːʃn]	n.执行，实行，履行
reuse	[ˌriːˈjuːz]	vt.复用，重新使用
inheritance	[inˈheritəns]	n.继承；遗传
principle	[ˈprinsəpl]	n.原则，原理；准则
multitasking	[ˌmʌltiˈtɑːskiŋ]	n.多（重）任务处理
bottom-up	[ˈbɒtəmʌp]	adj.自底向上
developer	[diˈveləpə]	n.开发者
recompilation	[ˌriːkɒmpiˈleiʃn]	n.重新编译，再编译
debug	[diːˈbʌg]	vt.调试程序，排除故障
immature	[ˌiməˈtjʊə]	adj.不成熟的；未完成的；粗糙的
standard	[ˈstændəd]	adj.标准的，合格的
macro	[ˈmækrəʊ]	n.宏观的，大的
coding	[ˈkəʊdiŋ]	n.编写代码，译码
compiler	[kəmˈpailə]	n.编译器，编译程序
configuration	[kənˌfigjuˈreiʃn]	n.配置；布局；构造
accommodate	[əˈkɒmədeit]	v.容纳；适应
dictate	[dikˈteit]	vt.命令，指示；控制，支配
cumbersome	[ˈkʌmbəsəm]	adj.缓慢复杂的，冗长的；麻烦的

Phrases

programming language	编程语言
be capable of ...	有……能力（或技能）的，能……的
object-oriented design	面向对象设计
be accustomed to ...	习惯于……
adjust to	调整，调节
be suitable for ...	适合……的
data hiding	数据隐藏
be appropriate for	适合于，适用于
multi-paradigm programming	多范式编程
Virtual Machine Technology	虚拟机技术
memory manager	内存管理器
be acquainted with	熟悉
pattern matching	模式匹配
automatic backtracking	自动回溯

Abbreviations

WORA (Once Written Read/Run Anywhere)　　一次写成/处处可用

Text B 参考译文
用于人工智能的编程语言

　　实施任何类型的 AI 系统都必定涉及计算机编码，并且有多种编程语言可用于特定的 AI 或机器学习任务。让我们看看有哪些编程语言将对你的特定用例最有利。

　　Python，Java，C ++，Lisp 和 Prolog 都是用于人工智能的主要 AI 编程语言，能够满足不同软件的开发和设计中的不同需求。它们对于给定的项目来说都有其各自的优点和缺点，因此，开发人员可以根据满足应用程序要求的功能和特性来选择具体的编程语言。

1. Python

　　Python 是迄今为止在人工智能中使用最广泛的编程语言，因为它具有易于学习的语法、庞大的库和框架、对众多 AI 算法的动态适用性，并且相对易于编写。

Python 支持多种面向种类，包括功能、面向对象和过程。此外，其庞大的社区有助于将该语言保持在计算机科学行业的最前沿。

优点：
- Python 具有丰富多样的库和工具。
- Python 支持算法测试，而无须实施它们。
- Python 支持面向对象的设计，这提高了程序员的生产率。
- 与 Java 和 C ++相比，Python 的开发速度更快。

缺点：
- 习惯于使用 Python 的开发人员在尝试使用其他语言进行 AI 编程时，难以适应完全不同的语法。
- 与 C ++和 Java 不同，Python 在解释器的帮助下工作，这使 AI 开发中的编译和执行速度变慢。
- 不适合移动计算。

2. C ++

C ++是最快的计算机语言，其速度可让对时间敏感的 AI 编程项目受益。它提供了更快的执行速度，并且响应时间更短，适用于搜索引擎和计算机游戏开发。此外，C ++允许广泛使用算法，并且可以有效地使用统计 AI 技术。另一个重要因素是，由于继承和数据隐藏，C ++支持在开发中重用程序，从而节省了时间和成本。C ++适用于机器学习和神经网络。

优点：
- C ++非常适合为复杂的 AI 问题找到解决方案。
- C ++具有丰富的库函数和编程工具集合。
- C ++是一种多范式编程，它支持面向对象的原则，因此对于实现有组织的数据很有用。

缺点：
- 多任务处理能力差；C ++仅适用于实现特定系统或算法的核心或基础部分。
- 它遵循自下而上的方法，因此非常复杂，这使得新手开发人员很难使用它来编写 AI 程序。

3. Java

Java 也是一种遵循面向对象原则和"一次可读取/可在任何地方运行"（WORA）原则的多范式语言。它是一种 AI 编程语言，可以在支持它的任何平台上运行，而无须重新编译。

Java 是最常用的一种编程语言。它不仅仅用于 AI 开发，其语法的主要部分从 C 和 C ++派生。Java 不仅适用于 NLP 和搜索算法，还适用于神经网络。

优点：

Java 非常易于移植；由于使用了虚拟机技术，因此很容易在不同平台上实现。
- 与 C++不同，Java 易于使用，甚至是调试。

Java 具有自动内存管理器，可简化开发人员的工作。

缺点：
- 但是，Java 比 C++慢，它的执行速度较低，响应时间更长。
- 尽管具有高度可移植性，但在较旧的平台上，Java 要求对软件和硬件进行重大更改以便于使用。
- Java 还是 AI 的一种普遍不成熟的编程语言，因为仍在进行一些开发。

4. Lisp

Lisp 是用于人工智能开发的另一种语言。它是计算机编程语言的一种，是仅次于 Fortran 的第二古老的编程语言。随着时间的推移，Lisp 已经发展成为强大、动态的编码语言。

由于能够提供给开发人员的自由，有些人认为 Lisp 是最好的 AI 编程语言。Lisp 之所以在 AI 中使用，是因为 Lisp 具有用于快速原型制作和实验的灵活性，这反过来又有助于 Lisp 成长为标准的 AI 语言。例如，Lisp 有一个独特的宏系统，可促进探索和实施不同级别的智能。

与大多数 AI 编程语言不同，由于 Lisp 可以适应开发人员正在编写的解决方案的需求，因此它在解决特定问题方面效率更高。它非常适合归纳逻辑项目和机器学习。

优点：
- Lisp 由编译器（而不是解释器）支持，因此可以快速高效地进行编码。
- 自动内存管理器是为 Lisp 发明的，因此具有垃圾回收功能。
- Lisp 提供对系统的特定控制，以最大程度地使用它们。

缺点：
- 很少有开发人员熟悉 Lisp 编程。
- 作为一种老式的编程语言，要把它用于人工智能，Lisp 需要配置新的软件和硬件。

5. Prolog

Prolog 也是最古老的编程语言之一，也适合于编程 AI 的开发。与 Lisp 一样，它也是人工智能的主要计算机语言。它具有灵活框架的机制，开发人员因而喜欢使用。它是一种基于规则的声明性语言，因为它包含决定其人工智能编码语言的事实和规则。

Prolog 支持基本机制，例如模式匹配、基于树的数据结构以及对 AI 编程必不可少的自动回溯。除了在 AI 项目中广泛使用之外，Prolog 还用于创建医疗系统。

优点：
- Prolog 具有内置列表处理，对于表示基于树的数据结构至关重要。
- Prolog 对于 AI 程序的快速原型制作非常有效。
- Prolog 允许在程序运行的同时创建数据库。

缺点：
- Prolog 很古老。
- 某些功能没有实现完全标准化，这使开发人员的工作较为烦琐。

6. 结论

就像大多数软件应用程序的开发一样，开发人员也在使用多种语言来编写人工智能项目，但是迄今为止还没有任何一种完美的编程语言是可以完全速配人工智能项目的。编程语言的选择往往取决于对人工智能应用程序的期望功能。

Exercises

[Ex. 1] Answer the following questions according to Text A.

1. Why isn't developing an AI solution a one dimensional project?
2. What capabilities should a robust AI development platform offer ideally?
3. What are the ML capabilities of the Azure AI platform mentioned in the passage?
4. What is form recogniser?
5. What AI models do the Azure cognitive services include?
6. Under what condition can you launch Google compute engine instances?
7. What can you do with Google Cloud AI's speech recognition capabilities?
8. What is a key advantage of IBM Watson?
9. What is Infosys Nia? What AI capabilities does it offer?
10. Where does Dialogflow run? What are the key capabilities offered by Dialogflow?

[Ex. 2] Answer the following questions according to Text B.

1. What are major AI programming language used for artificial intelligence capable of satisfying different needs in the development and designing of different software?
2. Why is Python by far the most popular programming language used in artificial intelligence today?
3. What are the drawbacks of Python?
4. What is C++? What are its advantages?

5. What is Java?
6. What are the advantages of Java?
7. Why is Lisp used in AI?
8. What are the advantages of Lisp?
9. What basic mechanisms does Prolog support?
10. Is there a perfect programming language that can fully match artificial intelligence projects so far? What does the choice of programming language often depend on?

[Ex. 3] Translate the following terms or phrases from English into Chinese and vice versa.

1. compute engine   1. _____
2. knowledge repository   2. _____
3. machine translation   3. _____
4. object recognition   4. _____
5. open source   5. _____
6. *vt.* 访问，存取   6. _____
7. *n.* 聊天机器人   7. _____
8. *vt.* 定制，客户化   8. _____
9. *n.* 构架，框架   9. _____
10. *n.* 样板；模板   10. _____

[Ex. 4] Translate the following passages into Chinese.

The Best Artificial Intelligence Programming Languages – Java vs. Python

1. How Do You Think Java Is Important for Artificial Intelligence

Java is considered as one of the best languages for AI projects. If we examine artificial intelligence programming, we'll find machine learning solutions, neural networks, search algorithms and multi-robot systems use Java.

We know how artificial intelligence is highly connected with algorithms and Java has this potential to code different types of algorithms. Java tools create appealing graphics and interfaces and have a great number of Java AI libraries. Some notable features that Java offer to AI are:

Easy to debug: If you are a Java developer, debugging must be one of your skills. Java applications are easily debugged and easy to use.

Easy-to-code algorithms and high performance: Java is faster than other traditional interpreted programming languages. Java facilitates and offers high performance with the use of just-in-time compiler.

Simplified work and large-scale projects: Java is effectively used for large scale project development. We can access the library and tools ecosystem in Java.

Intelligent product development: Java has a built-in garbage collector that automatically deletes useless data, facilitates visualization and incorporates Swing and Standard Widget Toolkit.

Versatile, transparent, and easy to maintain: Java Byte code can be carried on any platform and maintained easily.

Java has a big global community of developers: Java community has millions of members worldwide. If you are a beginner, you can easily learn and create effective solutions as there is always someone who can answer your questions. Free online tutorials also help in learning Java easily.

2. What Does Python Offer to Artificial Intelligence

We know that a developer can directly run a program written in Python. The best part of Python is undoubtedly its readability and simple syntax.

There are various benefits of choosing Python to develop AI projects.

Less code: We don't have to code algorithms using Python. Less coding saves time and makes the whole process easier.

A great library ecosystem: Python is the most popular programming language used for AI because of the pre-built libraries. You do not have to code from the very beginning every time using Python.

Good visualization options: Python libraries offer libraries like Matplotlib that allows us to build charts, histograms, and plots for effective visualization that can create clear reports. We can represent data using great visualization tools.

Readability: Python is very easy to read, change, copy or share.

Platform Independent: Python can run on multiple platforms including Windows, MacOS, Linux, Unix etc. We can implement several small-scale changes to transfer the process from one platform to another.

[Ex. 5] Fill in the blanks with the words given below.

| features | separate | cycle | parallel | interaction |
| relations | earlier | adopt | framework | phases |

SDLC (Software Development Life Cycle)

1. Why SDLC

Here are the prime reasons why SDLC is important for developing a software system.

- It offers a basis for project planning, scheduling, and estimating.
- It provides a ___1___ for a standard set of activities and deliverables.
- It is a mechanism for project tracking and control.
- IT increases visibility of project planning to all involved stakeholders of the development process.
- It increases and enhances development speed.
- It improves client ___2___.
- It helps you to decrease project risk and project management plan overhead.

2. Popular SDLC models

2.1 Waterfall model

The waterfall is a widely accepted SDLC model. In this approach, the whole process of the software development is divided into various ___3___. In this SDLC model, the outcome of one phase acts as the input for the next phase.

This SDLC model is documentation-intensive, with ___4___ phases documenting what need to be performed in the subsequent phases.

2.2 Incremental Approach

The incremental model is not a ___5___ model. It is essentially a series of waterfall cycles. The requirements are divided into groups at the start of the project. For each group, the SDLC model is followed to develop software. The SDLC process is repeated, with each release adding more functionality until all requirements are met. In this method, every ___6___ act as the maintenance phase for the previous software release. Modification to the incremental model allows development cycles to overlap. After that subsequent cycle may begin before the previous cycle is complete.

2.3 V-Model

In this type of SDLC model, the phase is planned in ___7___. So, there are verification phases on one side and the validation phase on the other side. V-Model joins by Coding phase.

2.4 Agile Model

Agile methodology is a practice which promotes to continue ___8___ of development and testing during the SDLC process of any project. In the agile method, the entire project is divided into small incremental builds. All of these builds are provided in iterations, and each iteration lasts from one to three weeks.

2.5 Spiral Model

The spiral model is a risk-driven process model. This SDLC model helps the team to ___9___ elements of one or more process models like a waterfall, incremental, waterfall, etc. This model adopts the best ___10___ of the prototyping model and the waterfall model. The spiral methodology is a combination of rapid prototyping and concurrency in design and development activities.

Unit 10
New Technology of Artificial Intelligence

Text A
Artificial Intelligence as a Service

扫码听课文

The concept "everything as a service" refers to any software that can be called upon across a network because it uses cloud computing. In most cases, the software is available off the shelf, meaning that you can buy it from a third-party vendor, make a few tweaks, and begin using it nearly immediately, even if it hasn't been totally customized to your system.

For companies that can't or are unwilling to build their own clouds or build, test, and utilize their own artificial intelligence systems, AI as a service is the solution. AI as a service provides the following functions:

- Allowing the company to focus on their core business, not becoming data and machine learning experts.
- Keeping costs transparent.
- Significantly lowering risk of investment.
- Increasing the benefits from data.
- Increasing strategic flexibility, because AI as a service is flexible and dynamic.

1. Types of AIaaS

If we understand AIaaS as artificial intelligence off the shelf, what are we getting when we purchase a service?

Well-known types of AIaaS include:

- Bots and digital assistance. These can include, for example, chat bots that use natural language processing (NLP) algorithms to learn from conversation with human beings and imitate the language patterns while providing answers. This frees up customer service employees to focus on more complicated tasks. These are the most widely used type of AIaaS right now.

- Cognitive computing APIs. Short for application programming interface, APIs are a way for developers to add a specific technology or service into the application they are building without writing the code from scratch. Common options for APIs include NLP, computer speech and computer vision, translation, knowledge mapping, search, and emotion detection.

- Machine learning frameworks. These are tools that developers can use to build their own model that learns over time from existing company data. Machine learning is often associated with big data but can have other uses and these frameworks provide a way to build in machine learning tasks without needing the big data environment.

- Fully-managed machine learning services. If machine learning frameworks are the first step towards machine learning, this option is a way to add in richer machine learning capabilities using templates, pre-built models, and drag-and-drop tools to assist developers in building a more customized machine learning framework.

2. Vendors of AIaaS

The four major vendors of AIaaS are well-known, including Amazon Web Services (AWS), Microsoft Azure, Google Cloud, and IBM Cloud. Each vendor offers different types of bots, APIs, and machine learning frameworks, and all except IBM are currently offering fully-managed machine learning options, too.

Other well-known technology firms are starting to get into the competition, including Oracle, BMC, and SalesForce.

There are countless start-ups who are focusing on various parts of AIaaS as well. As in all industries, it's not uncommon for the larger companies to purchase the smaller companies in order to add the developed services to their portfolios.

3. The Benefits and Drawbacks of AIaaS

Like other "as a service" offerings, AIaaS offers many of the same benefits, including:

- Advanced infrastructure at a fraction of the cost. Successful AI and machine learning often require many parallel machines and GPUs that must be speedy. Prior to AIaaS, a company may decide the initial investment and ongoing upkeep of physical and digital equipment too much. Now, AIaaS means companies can harness the power of machine learning at significantly lower costs. This means you can continue working on your core business, not training and spending on areas that partially support decision-making.

- Transparency. Hand in hand with lower costs, there's a lot of transparency within AIaaS: Pay for what you use. Though machine learning requires a lot of power when it's running, you may only need that power in short amounts of time — you don't have to be running AI non-stop.
- Usability. While many AI options are currently open-source, they aren't always user-friendly. This means your developers are spending time installing and developing the machine learning technology. AIaaS instead is ready out of the box, so developers can harness the power of AI without becoming experts first.
- Scalability. AIaaS allows you to start with smaller projects to learn if it's the right fit for certain projects. As you gain experience with your own data, you can tweak your service and scale up or down as project demands change.

The drawbacks of AIaaS are similar to other "as a service" offerings as well, including:

- Reduced security. AI and machine learning depend on significant amounts of data, which means your company must share that data with third-party vendors. Data storage and access and data transit ting to servers must be secured to ensure the data isn't improperly accessed, shared, or tampered with.
- Reliance. Because you're working with one or more third-parties, you're relying on them to provide the information you need. This isn't inherently a problem, but it can lead to lag time or other issues if any problems arise.
- Reduced transparency. In AIaaS, you're buying the service, but not the access. Some consider "as a service" offerings a black box — you know the input and the output, but you don't understand the inner-workings, like which algorithms are being used, whether the algorithms are updated, and which versions apply to which data. This may lead to confusion or miscommunication regarding the stability of your data or the output.
- Data governance. Particular industries, such as medical fields, and particular countries may limit whether or how data be stored in a cloud, which altogether may prohibit your company from taking advantage of certain types of AIaaS.
- Long-term costs. Costs can quickly spiral with "as a service" offerings, and AIaaS is no exception. As you wade deeper into AI and machine learning, you may be seeking out more complex offerings, which can cost more and require that you hire and train staff with more specific experience. As with anything, though, the costs may be a wise investment for your company.

As a rapidly emerging field, AI as a service has plenty of benefits that bring early adapters to the table as well as plenty of drawbacks that mean room for improvement. While there may be bumps in the road while being developed, AIaaS is likely to be as important, if not more so, than previous "as a service" offerings.

New Words

third-party	[ˈθɜːd ˈpɑːti]	adj.第三方的
tweak	[twiːk]	vt.稍稍调整（机器、系统）
utilize	[ˈjuːtɪlaɪz]	vt.利用，使用
option	[ˈɒpʃn]	n.选择；选项；选择权
transparent	[trænsˈpærənt]	adj.透明的；显而易见的
significantly	[sɪgˈnɪfɪkəntli]	adv.极大地；显著地；大幅度地
imitate	[ˈɪmɪteɪt]	vt.模仿，效仿
prebuilt	[preˈbɪlt]	adj.预建的
drag-and-drop	[ˈdræg ənˈdrɒp]	n.（鼠标的）拖放动作
countless	[ˈkaʊntlɪs]	adj.无数的
start-up	[ˈstɑːt ʌp]	n.启动
uncommon	[ʌnˈkɒmən]	adj.不寻常的；罕见的；非凡的
portfolio	[pɔːtˈfəʊliəʊ]	n.代表作品集
speedy	[ˈspiːdi]	adj.快的，迅速的；敏捷的
upkeep	[ˈʌpkiːp]	n.保养，维修；保养费，维修费
equipment	[ɪˈkwɪpmənt]	n.设备，装备；器材
harness	[ˈhɑːnɪs]	vt.利用；控制
transparency	[trænsˈpærənsi]	n.透明度；透明性
usability	[ˌjuːzəˈbɪləti]	n.可用性；适用性
user-friendly	[ˌjuːzə ˈfrendli]	adj.用户友好的
scalability	[ˌskeɪləˈbɪləti]	n.可伸缩性
demand	[dɪˈmɑːnd]	v.& n.需求，需要
transit	[ˈtrænzɪt]	vt.传输，运送
server	[ˈsɜːvə]	n.服务器
improperly	[ɪmˈprɒpəli]	adv.不正确地，不适当地
share	[ʃeə]	v.共享，分享
reliance	[rɪˈlaɪəns]	n.依靠，依赖；信任，信赖
inherently	[ɪnˈhɪərəntli]	adv.天性地，固有地
confusion	[kənˈfjuːʒn]	n.混淆；混乱；困惑

miscommunication	[ˈmisˌkəmju:niˈkeiʃən]	n.错误传达
stability	[stəˈbiləti]	n.稳定（性），稳固
prohibit	[prəˈhibit]	vt.禁止，阻止，防止；不准许
spiral	[ˈspaiərəl]	v.螺旋式的上升（或下降）
adapter	[əˈdæptə]	n.适应者；适配器
improvement	[imˈpru:vmənt]	n.改进，改善，改良

Phrases

everything as a service	一切即服务
cloud computing	云计算
off the shelf	现货供应，现成的
be applied to	适用于；应用于
digital assistance	数字助理，数字助手
chat bot	聊天机器人
free up	开放（市场等）；释放
cognitive computing	认知计算
from scratch	从头做起，从零开始
big data	大数据
a fraction of	一小部分
prior to ...	在……之前
initial investment	初期投资，初始投资
tamper with	损害；篡改
black box	黑箱，黑匣子
data governance	数据治理
emerging field	新型领域
plenty of	很多，大量的
bumps in the road	磕磕碰碰，诸多困难

Abbreviations

AIaaS (AI as a Service)	人工智能即服务
GPU (GraphicsProcessingUnit)	图形处理器

Text A 参考译文
人工智能即服务

"一切即服务"这一概念是指可以通过网络调用的任何软件,因为它使用云计算。在大多数情况下,该软件都是现成的,这意味着你可以从第三方供应商处购买该软件,进行一些调整,然后几乎立即就能使用它,即使该软件尚未完全针对你的系统进行定制。

对于无法或不愿意构建自己的云或者无法或不愿构建、测试和利用自己的人工智能系统的公司,AI 即服务是不错的解决方案。人工智能即服务提供如下功能:
- 让公司专注于核心业务,而不是成为数据和机器学习专家。
- 保持成本透明。
- 大大降低投资风险。
- 增加数据收益。
- 提高战略灵活性,因为 AI 即服务是灵活而动态的。

1. AIaaS 的类型

如果我们将 AIaaS 理解为现成的人工智能,那么在购买服务时会得到什么?
广为人知的 AIaaS 类型包括:
- 机器人和数字助理。例如聊天机器人,它们使用自然语言处理(NLP)算法,从与人类的对话中学习并在提供答案的同时模仿语言模式。这使客户服务人员可以腾出时间专注于更复杂的任务。这些是目前使用最广泛的 AIaaS 类型。
- 认知计算 API。API 是应用程序编程接口的简称,是开发人员无需从头开始编写代码就可将特定的技术或服务添加到正在构建的应用程序中的一种方法。API 的常见选项包括 NLP、计算机语音和计算机视觉、翻译、知识映射、搜索和情感检测。
- 机器学习框架。这些是开发人员可以用来构建自己的模型的工具,这些模型可以随着时间的推移从现有公司数据中学习。机器学习通常与大数据相关联,但还可以有其他用途——这些框架提供了一种无须大数据环境即可构建机器学习任务的方法。
- 完全托管的机器学习服务。如果机器学习框架是迈向机器学习的第一步,则此选项是一种使用模板,使用预构建模型和拖放工具来添加更为丰富的机器学习功能,以帮助开发人员构建客户化程度更高的机器学习框架。

2. AIaaS 的供应商

AIaaS 的四个主要供应商是 Amazon Web Services(AWS)、Microsoft Azure、Google Cloud 和 IBM Cloud。每个供应商都提供不同类型的机器人、API 和机器学习框架。除 IBM 外,所有

供应商目前都还提供完全托管的机器学习选项。

其他知名技术公司也开始参与竞争，包括 Oracle、BMC 和 SalesForce。

也有无数的初创公司专注于 AIaaS 的各个方面。与所有行业一样，大型公司购买小型公司以便将已开发的服务添加到自己的产品集中，这并不少见。

3. AIaaS 利弊

与其他"即服务"产品一样，AIaaS 也具有许多相同的优势，包括：

- 先进的基础设施，而成本却很小。成功的 AI 和机器学习通常需要许多并行机器和 GPU，它们必须运行快速。在使用 AIaaS 之前，公司可能会过多地投资于物理设备和数字设备的购买和维护。现在，AIaaS 意味着公司能够以更低的成本利用机器学习的力量。这意味着你可以继续从事核心业务，而不必在部分支持决策的领域进行培训和支出。
- 透明度。与更低的成本相伴，AIaaS 的透明度很高：为你使用的东西付费。尽管机器学习需要大量的动力才能运行，但是你可能只需要在短时间内获得这种动力——你不必不停歇地运行 AI。
- 可用性。尽管许多 AI 选项目前都是开源的，但它们并不总是用户友好的。这意味着你的开发人员正在花时间安装和开发机器学习技术。相反，AIaaS 可以立即使用，因此开发人员可以利用 AI 的力量而无须先成为专家。
- 可扩展性。AIaaS 允许你从较小的项目开始，以了解它是否适合某些项目。当你获得自己的数据经验时，就可以调整服务，并根据项目需求的变化来扩大或缩小规模。

AIaaS 的缺点也类似于其他"即服务"产品，包括：

- 降低了安全性。人工智能和机器学习取决于大量数据，这意味着你的公司必须与第三方供应商共享这些数据。必须保护数据的存储、访问和到服务器的数据传输，以确保不会对数据进行不正确的访问、共享或篡改。
- 具有依赖性。因为你正在与一个或多个第三方合作，所以你依靠它们来提供所需的信息。这本身并没有什么问题，但它可能会导致滞后或其他问题。
- 透明度降低。在 AIaaS 中，你购买的是服务，而不是访问权。有些人认为"即服务"提供了一个黑匣子——你知道输入和输出，但你不了解内部工作原理，例如正在使用哪种算法、是否更新算法以及哪些版本适用于哪种数据。这可能会导致你的数据或输出稳定性出现混乱或沟通不畅。
- 数据治理。特定行业（例如医疗领域）和特定国家/地区可能会限制将数据存储在云中或限制将其存储在云中的方式，甚至会完全禁止你的公司利用某些类型的 AIaaS。
- 长期成本。使用"即服务"产品可能使成本迅速攀升，而 AIaaS 也不例外。随着你更深入地学习 AI 和机器学习，你可能正在寻找更复杂的产品，这些产品可能会更昂贵，并且需要你雇用和培训具有特定经验的员工。尽管，就像任何事情一样，这种成本对于你的公司来说都是明智的投资。

作为一个快速发展的领域，"AI即服务"具有许多优势，可以使早期适应者浮出水面，同时也具有许多缺点，这些缺点意味着有改进的余地。虽然开发AIaaS可能会遇到困难，但它可能与以前的"即服务"产品一样重要（如果不是更重要多的话）。

Text B
The Role of Artificial Intelligence in Cloud Computing

1. Introduction

Cloud computing services have morphed from platforms such as Google App Engine and Azure to infrastructure which involves the provision of machines for computing and storage. In addition to this, cloud providers also offer data platform services which span the different available databases. This chain of development points to the direction of the growth of artificial intelligence and cloud computing.

扫码听课文

2. The Existing Types of Cloud Application Development Services

2.1 Infrastructure as a Service (IaaS)

This is the cloud App development service which is most employed by users. It allows you to pay based on the usage of the services provided, a truly flexible plan. The services provided include renting storage, networks, operating systems, servers and virtual machines.

2.2 Platform as a Service (PaaS)

This service is designed to make web creation and mobile App design easier by having an inbuilt infrastructure of servers, networks, databases and storage that eliminates the need to constantly update them or manage them.

2.3 Software as a Service (SaaS)

With this, the cloud provider and not the user is tasked with management and maintenance and all the user has to do to gain access is connected to the application over the internet with a web browser on his phone, tablet or PC. The SaaS is available over the internet on demand or on a subscription basis.

3. Types of Cloud Deployment

3.1 Public Cloud
For public clouds like the Microsoft Azure, the cloud provider owns and manages all hardware, software and other supporting infrastructures and is responsible for delivering computing resources — servers, storage — over the internet. As a user, you gain access to these services and manage your account through the web browser.

3.2 Private Cloud
Just as the name implies, a private cloud's services and infrastructure are maintained on a private network either by the providing company or a hired third-party service provider. It is used by a single organization and is sometimes located in the company's on-site data center itself.

3.3 Hybrid Cloud
This is a fusion of both the public and the private cloud services. How is this made possible? It is made available by the integration of the personalized data and applications shared by both platforms. Clients looking for more flexible cloud App development solutions and a wide range of deployment options are advised to embrace this technology.

4. Results from the Merger between Artificial Intelligence and Cloud Computing

4.1 Artificial Intelligence Infrastructure for Cloud Computing
We can generate machine learning models when a large set of data is applied to certain algorithms, and it becomes important to leverage the cloud for this. The models are able to learn from the different patterns which are gleaned from the available data.

As we provide more data for this model, the prediction gets better and the accuracy is improved. For instance, for ML models which identify tumors, thousands of radiology reports are used to train the system. This pattern can be used by any industry since it can be customized based on the project needs. The data is the required input and this comes in different forms — raw data, unstructured data, etc.

Because of the advanced computation techniques which require a combination of CPUs and GPUs, cloud providers now provide virtual machines with incredibly powerful GPUs. Also, machine learning tasks are now being automated using services which include batch processing, serverless computing, and orchestration of containers. IaaS also helps in handling predictive analytics.

4.2 Artificial Intelligence Services for Cloud Computing

Even without creating a unique ML model, it is possible to enjoy services provided by the AI systems. For instance, text analytics, speech, vision, and machine language translation are accessible to developers. They can simply integrate this into their development projects.

Although these services are generic and are not tailored to specific uses, cloud computing vendors are taking steps to ensure that this is constantly improved. Cognitive computing is a model which allows users to provide their personalized data which can be trained to deliver well-defined services. This way, the problem of finding the appropriate algorithm or the correct training model is eliminated.

5. Benefits of Leveraging Artificial Intelligence and Cloud Computing

5.1 Cost-Effectiveness

By being accessible through the internet, the cloud application development eliminates the need for expenses of on-site hardware and software purchases and setup. It also eliminates the need for on-site data centers and the expenses that come with it — IT experts to manage the centers, servers and round the clock electricity to power and cool the servers.

5.2 Increased Productivity

Unlike a hard drive or local storage device which requires a lot of IT management chores — hardware setup, software patching, racking and stacking, cloud computing is all internet based and as such has no need for this. This gives room for the IT team to focus on achieving other business goals.

5.3 Reliability

With a hard drive or physically accessible infrastructures, the risk of damage is heightened. One faces the risk of the crash, lost files, backup failure and so much more. However, cloud computing solutions ensure business continuity, faster and easier disaster recovery and easier data backup.

5.4 Availability of Advanced Infrastructure

AI applications are generally of high performance on servers with multiple and very fast GPUs. These systems are extremely expensive and unaffordable for many organizations. AI as a service in cloud application development becomes accessible to these organizations at a more affordable price.

6. Conclusion

The top cloud computing companies listed on GoodFirms strongly believe that the fusion of cloud computing services and AI technology will bring a significant change to the technology industry. Public cloud providers keep on investing in the growth of AI and this will continue to attract the right set of clients to this technology. Even though the technology is still in its early stage, the evolution to come is inevitable and we can expect phenomenal advancements in the future.

New Words

morph	[mɔ:f]	n.变种；变体
		vt.在屏幕上变换图像；改变
chain	[tʃein]	n.链，链条
		vt.链接
constantly	[ˈkɒnstəntli]	adv.不断地，时常地；始终，一直
internet	[ˈintənet]	n.互联网
browser	[ˈbraʊzə]	n.浏览程序；浏览器
subscription	[səbˈskripʃn]	n.订阅费；会员费
on-site	[ˈɒnˈsait]	adj.现场的
client	[ˈklaiənt]	n.顾客；客户端
leverage	[ˈli:vəridʒ]	n.杠杆作用，影响力
		v.发挥杠杆作用
radiology	[ˌreidiˈɒlədʒi]	n.放射学，辐射学
orchestration	[ˌɔ:kiˈstreiʃn]	n.编排，编曲
container	[kənˈteinə]	n.容器
cost-effectiveness	[ˌkɒstiˈfektivnis]	n.成本效用，成本效益
chore	[tʃɔ:]	n.零星工作；令人讨厌的或繁重的工作
setup	[ˈsetʌp]	n.设置
patch	[pætʃ]	n.补丁
		vt.修补
		vi.打补丁
rack	[ræk]	n.支架

crash	[kræʃ]	n.崩溃
failure	[ˈfeiljə]	n.失败
disaster	[diˈzɑ:stə]	n.灾难
unaffordable	[ˌʌnəˈfɔ:dəbl]	adj.买不起的，负担不起的
affordable	[əˈfɔ:dəbl]	adj.付得起的
fusion	[ˈfju:ʒn]	n.融合
attract	[əˈtrækt]	vt.吸引；引起……的好感（或兴趣） vi.具有吸引力；引人注意
evolution	[ˌi:vəˈlu:ʃn]	n.演变；进化；发展
inevitable	[inˈevitəbl]	adj.不可避免的；必然发生的
phenomenal	[fəˈnɒminl]	adj.显著的；异常的；非凡的，惊人的
advancement	[ədˈvɑ:nsmənt]	n.前进，进步；提升，升级

Phrases

operating system	操作系统
be tasked with	承担任务
cloud provider	云提供商
be connected to	与……有联系，与……有关联；与……连接
web browser	网络浏览器
cloud deployment	云部署
public cloud	公共云
private cloud	私有云，专用云
hybrid cloud	混合云
personalized data	个性化数据
raw data	原始数据
batch processing	批处理
text analytic	文本分析
local storage device	本地存储设备
hard drive	硬盘驱动器
data backup	数据备份

Abbreviations

IaaS (Infrastructure as a Service)　　基础设施即服务
PaaS (Platform as a Service)　　　　平台即服务
SaaS (Software as a Service)　　　　软件即服务
PC (Personal Computer)　　　　　　个人计算机
IT (Information Technology)　　　　信息技术

Text B 参考译文
人工智能在云计算中的作用

1. 引言

云计算服务已从 Google App Engine 和 Azure 之类的平台转变为基础设施，其中包括为计算和存储提供机器。除此之外，云提供商还提供跨不同现有数据库的数据平台服务。这条发展链指向了人工智能和云计算增长的方向。

2. 现有的云应用程序开发服务类型

2.1 基础架构即服务（IaaS）

这是用户使用最多的云应用开发服务，它使你能够根据所提供服务的使用情况付费，这是一个真正灵活的计划。提供的服务包括租赁存储、网络、操作系统、服务器和虚拟机（VM）。

2.2 平台即服务（PaaS）

该服务旨在通过拥有内置的服务器、网络、数据库和存储基础架构来简化网站创建和移动应用设计，而无须持续更新或管理它们。

2.3 软件即服务（SaaS）

管理和维护是云提供商而不是用户的任务，用户访问必须做的事情就是通过互联网用自己的手机、平板电脑或 PC 上的 Web 浏览器连接到应用程序。SaaS 可以按需或通过订阅在互联网上使用。

3. 云部署的类型

3.1 公共云

对于 Microsoft Azure 这样的公共云，云提供商拥有并管理所有硬件、软件和其他支持基础结构，并负责通过互联网交付计算资源（服务器、存储设备）。作为用户，你可以通过网络浏览器访问这些服务并管理账户。

3.2 私有云

顾名思义，私有云的服务和基础架构是由提供公司或雇用的第三方服务提供商在私有网络上维护的。它由单个组织使用，有时位于公司的现场数据中心。

3.3 混合云

这是公共云和私有云服务的融合，如何实现？通过将两个平台共享的个性化数据和应用程序集成在一起就可以。那些寻求更灵活的云应用程序开发解决方案和广泛部署选项的客户，建议采用此技术。

4. 人工智能与云计算合并的结果

4.1 用于云计算的 Artificial Intelligence 基础架构

当将大量数据应用于某些算法时，我们可以生成机器学习（ML）模型，为此利用云非常重要。这些模型能够从现有数据中收集的不同模式中学习。

当我们为该模型提供更多数据时，预测会变得更好，准确性也会提高。例如，对于识别肿瘤的 ML 模型，成千上万的放射学报告被用来训练系统。该模式可以应用于任何行业，因为它可以根据项目需求进行定制。必需输入数据，它会以不同的形式出现——原始数据、非结构化数据等。

由于需要结合使用 CPU 和 GPU 的先进计算技术，云提供商现在为虚拟机提供了功能强大的 GPU。而且，机器学习任务现在正在使用服务进行自动化，该服务包括批处理、无服务器计算和内容编排。IaaS 还有助于处理预测分析。

4.2 用于云计算的 Artificial Intelligence 服务

即使没有创建特别的 ML 模型，人们也可以享受 AI 系统提供的服务。例如，开发人员可以访问文本分析、语音、视觉和机器语言翻译，他们可以将其简单地集成到他们的开发项目中。

尽管这些服务是通用的，并且未针对特定用途进行了量身定制，但云计算供应商仍在采取步骤以确保其不断得到改进。认知计算是一种模型，它允许用户提供他们的个性化数据，可以对数据进行训练以提供明确的服务。这样，消除了寻找合适算法或正确训练模型的问题。

5. 利用 Artificial Intelligence 和云计算的好处

5.1 节约成本

可通过互联网进行访问，云应用程序开发不需要现场硬件和软件购买与设置费用。它还消除了对现场数据中心的需求以及随之而来的费用——由 IT 专家管理中心和服务器并且全天候为服务器供电和散热。

5.2 提高生产力

与需要大量 IT 管理工作的硬盘驱动器或本地存储设备（硬件设置、软件补丁、机架和堆栈）不同，云计算全部基于互联网，因此不需要这样做。这为 IT 团队提供了专注于实现其他业务目标的空间。

5.3 增强可靠

使用硬盘驱动器或可访问的基础结构，损坏的风险会增加，会有崩溃、文件丢失、备份失败等风险。但是，云计算解决方案可确保业务连续性，更快、更轻松的灾难恢复以及更容易的数据备份。

5.4 先进的基础设施变得触手可及

在具有多个且非常快速的 GPU 的服务器上使用 AI 应用程序时，通常性能很高。对于许多组织而言，这些系统非常昂贵负担不起，现在这些组织可以更实惠的价格使用云应用程序开发中的 AI 即服务。

6. 结论

在 GoodFirms 上排名靠前的顶级云计算公司坚信，云计算服务和 AI 技术的融合将为技术行业带来重大变化。公共云提供商继续投资于 AI 的增长，这将继续吸引适合该技术的客户群。尽管该技术仍处于早期阶段，但它的发展势不可当，我们可以预见未来会有惊人的进步。

Exercises

[Ex. 1] Answer the following questions according to Text A.

1. What does the concept "everything as a service" refer to?
2. What functions does AI as a service provide?
3. What are the well-known types of AIaaS?
4. What are APIs?
5. What are the four major vendors of AIaaS?
6. What benefits does AIaaS offer?
7. What does successful AI and machine learning often require?

8. What can you do as you gain experience with your own data?
9. What are the drawbacks of AIaaS?
10. What do AI and machine learning depend on? What does it mean? What must be done to ensure the data isn't improperly accessed, shared, or tampered with?

[Ex. 2] Fill in the following blanks according to Text B.
1. Cloud computing services have morphed from _____ such as Google App Engine and Azure to _____ which involves the provision of machines for _____.
2. The existing types of cloud application development services are _____, _____ and _____.
3. PaaS is designed to make web creation and _____ easier by having an inbuilt infrastructure of _____, _____, databases and storage that eliminates the need to constantly _____ or manage them.
4. There are three types of cloud deployment mentioned in the passage. They are _____, _____ and _____.
5. Just as the name implies, a private cloud's services and infrastructure are maintained on _____ either by the providing company or _____. It is used by _____ and is sometimes located in _____.
6. Because of the advanced computation techniques which require a combination of _____ and _____, cloud providers now provide virtual machines with incredibly powerful GPUs. Also, machine learning tasks are now being automated using services which include _____, _____ and _____. IaaS also helps in handling _____.
7. Cognitive computing is _____ which allows users to provide _____ which can be trained to deliver well-defined services. This way, the problem of _____ or _____ is eliminated.
8. The benefits of leveraging AI and cloud computing are _____, _____, _____ and _____.
9. With a hard drive or physically accessible infrastructures, the risk of damage is _____. One faces the risk of _____, _____, _____ and so much more. However, cloud computing solutions ensure _____, faster and easier _____ and easier data backup.
10. The top cloud computing companies listed on _____ strongly believe that the fusion of _____ and _____ will bring a significant change to the technology industry.

[Ex. 3] Translate the following terms or phrases from English into Chinese and vice versa.

1. be applied to   1. _____
2. cloud computing   2. _____
3. data governance   3. _____
4. everything as a service   4. _____
5. cognitive computing   5. _____
6. *n.*混淆；混乱；困惑   6. _____
7. *n.*设备，装备；器材   7. _____
8. *n.*改进，改善，改良   8. _____
9. *n.*服务器   9. _____
10. *adj.*用户友好的   10. _____

[Ex. 4] Translate the following passages into Chinese.

Artificial Intelligence and Cloud

1. Powering a Self-Managing Cloud with Artificial Intelligence

Artificial intelligence is being embedded into IT infrastructure to help streamline workloads and automate repetitive tasks. Some have gone as far as predicting that as AI becomes more sophisticated, private and public cloud instances will rely on these AI tools to monitor, manage, and even self-heal when an issue occurs. Initially, AI can be used to automate core workflows and then, over time, analytical capabilities can create better processes. Routine processes can be managed by the system itself, further helping IT teams capture the efficiencies of cloud computing and allowing them to focus on higher-value strategic activities.

2. Improving Data Management with Artificial Intelligence

At the cloud level, artificial intelligence tools are also improving data management. Consider the vast repositories of data that today's businesses generate and collect, as well as the process of simply managing that infrastructure — identifying data, ingesting it, cataloging it. Cloud computing solutions are already using AI tools to help with specific aspects of the data process. In banking, for example, even the smallest financial organization may need to monitor thousands of transactions per day.

AI tools can help streamline the way data is ingested, updated, and managed, so financial institutions can more easily offer accurate real-time data to clients. The same process can also help flag fraudulent activity or identify other areas of risk. Similar improvements can have a major impact on areas such as marketing, customer service, and supply chain data management.

3. Getting More Done with AI–SaaS Integration

Artificial intelligence tools are also being rolled out as part of larger Software-as-a-Service (SaaS) platforms to deliver more value. Increasingly, SaaS providers are embedding AI tools into their larger software suites to offer greater functionality and value to end users. Let's explore one popular example: the customer relationship management platform Salesforce and its Einstein AI tool. Salesforce introduced Einstein to help turn data into actionable insights businesses can use to sell more, improve their sales strategies, and engage with customers. The tools can help a business look for patterns in customer interactions, for example, to help advise sales on what method — like phone, email, or an in-person meeting — is more likely to drive a conversion. It can also be used to make "next step" recommendations based on the buying signals the tool is perceiving.

4. Utilizing Dynamic Cloud Services

AI as a service is also changing the ways businesses rely on tools. Consider a cloud-based retail module that makes it easier for brands to sell their products. The module has a pricing feature that can automatically adjust the pricing on a given product to account for issues such as demand, inventory levels, competitor sales, and market trends. An AI-powered pricing module such as this ensures that a company's pricing will always be optimized. It's not just about making better use of data, it's conducting that analysis and then putting it into action without the need for human intervention.

[Ex. 5] Fill in the blanks with the words given below.

| abilities | ideas | recognize | absorb | interact |
| intelligent | successful | agrees | sensors | elements |

Artificial intelligence is arguably the most exciting field in robotics. It's certainly the most controversial: Everybody ___1___ that a robot can work in an assembly line, but there's no consensus on whether a robot can ever be ___2___.

Like the term "robot" itself, artificial intelligence is hard to define. UltimateAI would be a recreation of the human thought process — a man-made machine with our intellectual ___3___. This would include the ability to learn just about anything, the ability to reason, the ability to use language and the ability to formulate original ___4___. Roboticists are nowhere near achieving this level of artificial intelligence, but they have made a lot of progress with more limited AI. Today's AI machines can replicate some specific ___5___ of intellectual ability.

Computers can already solve problems in limited realms. The basic idea of AI problem-solving is very simple, though its execution is complicated. Firstly, the AI robot or computer gathers facts about a situation through ___6___ or human input. The computer compares this information with stored data and decides what the information signifies. The computer runs through various possible actions and predicts which action will be most ___7___ based on the collected information. Of course, the computer can only solve problems it's programmed to solve — it doesn't have any generalized analytical ability. Chess computers are one example of this sort of machine.

Some modern robots also have the ability to learn in a limited capacity. Learning robots ___8___ if a certain action (moving its legs in a certain way, for instance) achieved a desired result (navigating an obstacle). The robot stores this information and attempts the successful action the next time it encounters the same situation. Again, modern computers can only do this in very limited situations. They can't ___9___ any sort of information like a human can. Some robots can learn by mimicking human actions. In Japan, roboticists have taught a robot to dance by demonstrating the moves themselves.

Some robots can ___10___ socially. Kismet, a robot at M.I.T's Artificial Intelligence Lab, recognizes human body language and voice inflection and responds appropriately. Kismet's creators are interested in how humans and babies interact, based only on tone of speech and visual cue. This low-level interaction could be the foundation of a human-like learning system.